T0388153

Clinical Management of Pediatric COVID-19

Clinical Management of Pediatric COVID-19

An International Perspective and Practical Guide

Edited by

Char Leung

School of Clinical Medicine, University of Cambridge, Cambridge, United Kingdom

Academic Press is an imprint of Elsevier
125 London Wall, London EC2Y 5AS, United Kingdom
525 B Street, Suite 1650, San Diego, CA 92101, United States
50 Hampshire Street, 5th Floor, Cambridge, MA 02139, United States
The Boulevard, Langford Lane, Kidlington, Oxford OX5 1GB, United Kingdom

ISBN: 978-0-323-95059-6

For information on all Academic Press publications visit our website at https://www.elsevier.com/books-and-journals

Publisher: Stacy Masucci
Acquisitions Editor: Patricia M. Osborn
Editorial Project Manager: Pat Gonzalez
Production Project Manager: Sajana Devasi P K
Cover Designer: Matthew Limbert

Typeset by TNQ Technologies

Dedication

To my mother Alice
Thank you for your genes.

Contents

7. COVID-19 vaccinations for children and adolescents

Katrina Nicolopoulos, Ketaki Sharma, Lucy Deng and Archana Koirala

8. Long COVID in children

Joseph L. Mathew and Kamal Kumar Singhal

Contributors

U. Allen, Division of Infectious Diseases, The Hospital for Sick Children, University of Toronto, Toronto, ON, Canada

Ioana Roxana Bordea, Department of Oral Rehabilitation, Faculty of Dentistry, Iuliu Hațieganu University of Medicine and Pharmacy, Cluj-Napoca, Romania

Loredana Capozzi, Istituto Zooprofilattico Sperimentale Della Puglia e della Basilicata, Foggia, Italy

Maria Contaldo, Multidisciplinary Department of Medical-Surgical and Dental Specialties, University of Campania Luigi Vanvitelli, Naples, Italy

Maria Franca Coscia, Department of Basical Medical Sciences, Neurosciences and Sense Organs (S.M.B.N.O.S.), University of Medicine Aldo Moro, Bari, Italy

Lucy Deng, National Centre for Immunisation and Research, Westmead, NSW, Australia; The University of Sydney, Sydney, NSW, Australia

Marina Di Domenico, Department of Precision Medicine, University of Campania Luigi Vanvitelli, Naples, Italy

Gianna Dipalma, Department of Interdisciplinary Medicine, University of Medicine Aldo Moro, Bari, Italy

Maria Teresa D'Oria, Department of Interdisciplinary Medicine, University of Medicine Aldo Moro, Bari, Italy; Department of Medical and Biological Sciences, University of Udine, Udine, Italy

Marco Farronato, UOC Maxillo-facial Surgery and Dentistry, Department of Biomedical, Surgical and Dental Sciences, School of Dentistry, Fondazione IRCCS Ca Granda, Ospedale Maggiore Policlinico, University of Milan, Milan, Italy

Mariantonietta Francavilla, Unit of Pediatric Imaging, Giovanni XXIII Hospital, Bari

Delia Giovanniello, Department of Toracic Surgery, Hospital "San Camillo Forlanini", Rome, Italy

H.E. Groves, Wellcome-Wolfson Institute for Experimental Medicine, Queens University, Belfast, Northern Ireland, United Kingdom

Jessica Gulliver, Department of Pathology and Laboratory Medicine, University of Wisconsin School of Medicine and Public Health, Madison, WI, United States; Department of Pathology and Laboratory Medicine, Ann & Robert H. Lurie Children's Hospital of Chicago, Chicago, IL, United States

Alessio Danilo Inchingolo, Department of Interdisciplinary Medicine, University of Medicine Aldo Moro, Bari, Italy

Francesco Inchingolo, Department of Interdisciplinary Medicine, University of Medicine Aldo Moro, Bari, Italy

Angelo Michele Inchingolo, Department of Interdisciplinary Medicine, University of Medicine Aldo Moro, Bari, Italy

Archana Koirala, National Centre for Immunisation and Research, Westmead, NSW, Australia; The University of Sydney, Sydney, NSW, Australia; Nepean Hospital, Kingswood, NSW, Australia

Ketan Kumar, Department of Pediatrics, Advanced Pediatrics Centre, Postgraduate Institute of Medical Education and Research (PGIMER), Chandigarh, Punjab, India

Char Leung, School of Clinical Medicine, University of Cambridge, Cambridge, United Kingdom; Department of Health Sciences, University of Leicester, Leicester, United Kingdom

Felice Lorusso, Department of Innovative Technologies in Medicine and Dentistry, University of Chieti-Pescara, Chieti, Italy

Maria Elena Maggiore, Department of Interdisciplinary Medicine, University of Medicine Aldo Moro, Bari, Italy

Giuseppina Malcangi, Department of Interdisciplinary Medicine, University of Medicine Aldo Moro, Bari, Italy

Antonio Mancini, Department of Interdisciplinary Medicine, University of Medicine Aldo Moro, Bari, Italy

Grazia Marinelli, Department of Interdisciplinary Medicine, University of Medicine Aldo Moro, Bari, Italy

Joseph L. Mathew, Department of Pediatrics, Advanced Pediatrics Centre, Post-graduate Institute of Medical Education and Research (PGIMER), Chandigarh, Punjab, India

Eduard Matkovic, Department of Pathology and Laboratory Medicine, University of Wisconsin School of Medicine and Public Health, Madison, WI, United States

S.K. Morris, Division of Infectious Diseases, The Hospital for Sick Children, University of Toronto, Toronto, ON, Canada

Damiano Nemore, Department of Interdisciplinary Medicine, University of Medicine Aldo Moro, Bari, Italy

Katrina Nicolopoulos, National Centre for Immunisation and Research, Westmead, NSW, Australia; The University of Sydney, Sydney, NSW, Australia

Ludovica Nucci, Multidisciplinary Department of Medical-Surgical and Dental Specialties, University of Campania Luigi Vanvitelli, Naples, Italy

Antonio Parisi, Istituto Zooprofilattico Sperimentale Della Puglia e della Basilicata, Foggia, Italy

Biagio Rapone, Department of Interdisciplinary Medicine, University of Medicine Aldo Moro, Bari, Italy

Roberto M. Richheimer-Wohlmuth, Cardiology Department at the Cardiovascular Division of the American British Cowdray Medical Center, Mexico City, Mexico; Paediatric Department of the American British Cowdray Medical Center, Mexico City, Mexico

Luigi Santacroce, Department of Interdisciplinary Medicine, University of Medicine Aldo Moro, Bari, Italy

Antonio Scarano, Department of Innovative Technologies in Medicine and Dentistry, University of Chieti-Pescara, Chieti, Italy

Arnaldo Scardapane, Department of Interdisciplinary Medicine, University of Medicine Aldo Moro, Bari, Italy

Rosario Serpico, Multidisciplinary Department of Medical-Surgical and Dental Specialties, University of Campania Luigi Vanvitelli, Naples, Italy

Ketaki Sharma, National Centre for Immunisation and Research, Westmead, NSW, Australia; The University of Sydney, Sydney, NSW, Australia

Lilia M. Sierra-Galan, Cardiology Department at the Cardiovascular Division of the American British Cowdray Medical Center, Mexico City, Mexico

Kamal Kumar Singhal, Department of Pediatrics, Advanced Pediatrics Centre, Postgraduate Institute of Medical Education and Research (PGIMER), Chandigarh, India

Gianluca Martino Tartaglia, UOC Maxillo-facial Surgery and Dentistry, Department of Biomedical, Surgical and Dental Sciences, School of Dentistry, Fondazione IRCCS Ca Granda, Ospedale Maggiore Policlinico, University of Milan, Milan, Italy

Luigi Vimercati, Department of Interdisciplinary Medicine, University of Medicine Aldo Moro, Bari, Italy

Edit Xhajanka, Department of Dental Prosthesis, Medical University of Tirana, Rruga e Dibrës, U.M.T., Tirana, Albania

Preface

By the time I am editing this book, it has already been more than 2 years since the outbreak of COVID-19. The pandemic has not ended, and new variants continue to emerge. A large body of studies has been done—from virology and epidemiological research to therapeutics. We have learned that the elderly and those with certain comorbidities are at high risk of SARS-CoV-2 infection and morbidity. We have also learned that the underage has a milder clinical course compared to adults. Consequently, we should ask ourselves, are pediatrics with COVID-19 overlooked because children are generally believed to fare better?

There are already several books covering different aspects of COVID-19. While some of them may have addressed children with COVID-19, there is a need to have a piece of work entirely dedicated to pediatric COVID-19—a book that highlights key findings and serves as a reference for clinicians and medical students in pediatrics. I hope this book satisfies this need.

Gathering contributors for this book was not an easy task at all as most of us including academics and clinicians are busy during the pandemic. This also means that getting the drafts done on the scheduled time was also problematic. Luckily, with Pat's help, Elsevier provided multiple platforms for manuscript submission, streamlining the process and making my life easier.

Each chapter is aimed to be a review study of a specific topic. Although a single title cannot cover all existing literature, our contributors carefully selected studies and presented key findings of clinical importance and presented. A wide range of topics on pediatric COVID-19 is covered. The first two chapters are set to provide background materials of the disease. The virology and epidemiology of SARS-CoV-2 such as the origin of the disease, route of transmission, and key epidemiologic metrics are discussed in Chapter 1. The pathogenesis of SARS-CoV-2 is discussed in Chapter 2 with a focus on the host–pathogen interaction, including tissue tropism, viral replication, and mechanism of dissemination, and how these factors interact in different age groups.

Chapters 3–6 focus on the clinical aspects of pediatric COVID-19. Chapter 3 is more descriptive with the aim to highlight signs and symptoms commonly seen in pediatrics. COVID-19-related complications such as multisystemic inflammatory syndrome in children are discussed in Chapter 4. Diagnosis of COVID-19 including laboratory tests and imaging is addressed in Chapter 5. Existing treatments are then discussed in Chapter 6.

The last two chapters focus on other topics—immunization and long COVID-19 in Chapters 7 and 8, respectively.

Lastly, I would like to thank all contributors for their effort, Pat Gonzalez of Elsevier for managing the book project and Stacy Masucci of Elsevier for inviting me to be the editor of the book. Without them, this book can never be published. I would also like to thank our reviewers Sarah Messiah, Luyu Xie, and Olivia Kapera of the University of Texas Health Science Center, Arthur Vengesai of Midlands State University, and other anonymous reviewers for their helpful and enlightening comments. Last but not least, I would like to thank my sister, Davily Leung, a graphic designer, for the cover image.

Char Leung

Chapter 1

Epidemiology and virology of SARS-CoV-2

Char Leung[1,2]
[1]*School of Clinical Medicine, University of Cambridge, Cambridge, United Kingdom;*
[2]*Department of Health Sciences, University of Leicester, Leicester, United Kingdom*

Introduction

Commonly known as COVID-19, the coronavirus disease 2019 is a viral respiratory disease caused by severe acute respiratory syndrome–associated coronavirus 2 (SARS-CoV-2), formerly known as 2019-nCoV. It was first identified in Wuhan, the capital of Hubei province in China, in December 2019 when a series of pneumonia cases were reported. The early phase of the outbreak in Wuhan is not well understood. Existing literature indicates that the first patient became symptomatic on December 1, 2019 [1]. A day before the closedown of the Market, the Wuhan Health Commission confirmed 27 cases of COVID-19 on December 31, 2019 [2]. According to the Chinese Center for Disease Control and Prevention, 33 environmental samples collected in the Huanan Seafood Wholesale Market in January 2020 were polymerase chain reaction (PCR) positive for SARS-CoV-2. In particular, 94% (31/33) of these samples were collected in the area where wildlife animals were heavily traded [3], suggesting the Market as the origin of SARS-CoV-2 [4]. The daily release of COVID-19 cases by Chinese health authorities did not start until January 10, 2020 when the number of cases increased to 41. The first case of COVID-19 outside China was reported on January 13, 2020, in Thailand where a 61-year-old woman traveling from Wuhan was detected with fever at the Suvarnabhumi Airport in Bangkok on January 8, 2020. As of end of January, more than 10,000 COVID-19 cases were confirmed in Wuhan, and the virus has spread to 22 countries. Meanwhile, the World Health Organization (WHO) declared the COVID-19 pandemic the sixth Public Health Emergency of International Concern.

Clinical Management of Pediatric COVID-19. https://doi.org/10.1016/B978-0-323-95059-6.00008-5

Virological characteristics of human coronaviruses

Commonly known as coronaviruses, *Cornidovirineae* or *Coronaviridae,* is a group of enveloped positive-strand RNA viruses. The two terms refer to the suborder and family in virus taxonomy, respectively, according to the International Committee on Taxonomy of Viruses (ICTV). Because *Coronaviridae* is the only family under *Cornidovirineae* as of 2021, the two terms have been used interchangeably. There are currently a total of 46 species of virus under *Cornidovirineae*, of which 19 and 10 have bats and birds as reservoir/host, respectively [5–8].

SARS-CoV-2 is one of the seven coronaviruses pathogenic to humans, the others being SARS-CoV (or SARS-CoV-1), MERS-CoV, OC43, NL63, 229E, and HKU1. According to the taxonomy by the ICTV, these viruses fall under the family of *Coronaviridae* that consists of the subfamilies *Letovirinae* and *Orthocoronavirinae*. Viruses of the latter can be further divided into 4 genera; *Alphacoronavirus*, *Betacoronavirus*, *Deltacoronavirus*, and *Gammacoronavirus*. Viruses of the latter two genera are not pathogenic to humans. NL63 and 229E are *Alphacoronaviruses,* whereas OC43, HKU1, MERS-CoV, SARS-CoV, and SARS-CoV-2 are *Betacoronavirus*.

Coronaviruses cause common colds

Human coronaviruses that produce mild upper respiratory diseases are OC43 first identified in 1965 in the United Kingdom [9], 229E in 1966 in the United States [10], NL63 in 2004 in the Netherlands [11], and HKU1 in 2005 in Hong Kong [12]. It has been proposed that NL63 and 229E originate from bat reservoirs, whereas OC43 and HKU1 are more likely to have speciated from rodent-associated viruses [13]. These coronaviruses constitute the second most common causative agent of all common colds, accounting for 10%–15%, after rhinoviruses [14]. It has been estimated that approximately 1 in 10 hospitalized children with respiratory tract infection is infected with at least one of these coronaviruses [15], the more recently discovered ones NL63 and HKU1 in particular [16,17]. The prevalence of infections among children is lowest in early summer, and the seasonal patterns can vary over time and between countries [18]. Studies have shown that many children and infants have been exposed to these strains although their seroprevalence varies over time and geographically [19–23].

Coronaviruses causing acute respiratory distress syndrome

Similar to SARS-CoV-2, SARS-CoV, and MERS-CoV can cause acute respiratory distress syndrome (ARDS). SARS-CoV emerged in Shunde of Foshan in China in November 2002 [24] and was first identified in Hong Kong in March 2003 [25]. The virus was believed to originate from bats and transmitted to humans through civets that were consumed as food. The data of 1425 SARS-CoV cases in Hong Kong suggest a case-fatality rate of 13% for patients younger than 60 years of age but 43% for those aged 60 years or above, In addition, approximately 25% of patients with SARS-CoV developed severe respiratory failure [26]. Children generally had a milder clinical course, and no deaths were reported [27]. MERS-CoV emerged in a hospital in Jordan in April 2012 [28] and was first isolated from a 60-year-old Saudi Arabian man admitted to the hospital in June 2012 [29]. While the virus is believed to originate from bats, patients could be infected by consuming and have close contact with camels infected with the virus. As the deadliest human coronavirus, the case-fatality rate of MERS-CoV is estimated to be 33%, based on 2562 confirmed cases [30]. Children with MERS-CoV appeared to have a lower mortality rate than adults [31] and were less likely to have a severe clinical course [32]. The phylogenetic tree of these viruses is shown in Fig. 1.1.

Severe acute respiratory syndrome–associated coronavirus 2

The structure of SARS-CoV-2 is shown in Fig. 1.2 and the genome organization in Fig. 1.3. Under the electronic microscope, the virus particle (also called the virion) is spherical in shape with spikes on the surface and has a diameter between 60 and 140 nm [33]. Each spike has a length of about 9–12 nm and is a glycosylated protein with two subunits, namely S1 and S2.

The S1 subunit forms the top part of the spike, largely consisting of the amino-terminal domain and the receptor binding domain (RBD) that attache

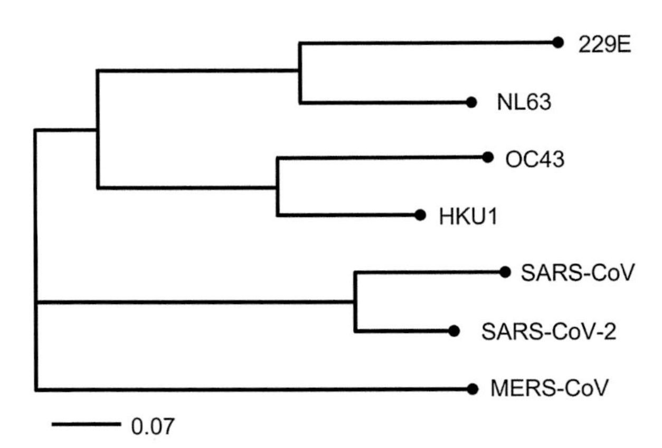

FIGURE 1.1 Phylogenetic tree of human coronaviruses (The node of each branch indicates the putative ancestor of the viruses, OC43 and HKU1, for example. The scale bar labeled "0.07" refers to the length of the horizontal line that represents the amount of genetic change, measured by the number of substitutions per nucleotide position. It is clear that severe acute respiratory syndrome–associated coronavirus 2 (SARS-CoV-2) is genetically close to SARS-CoV.).

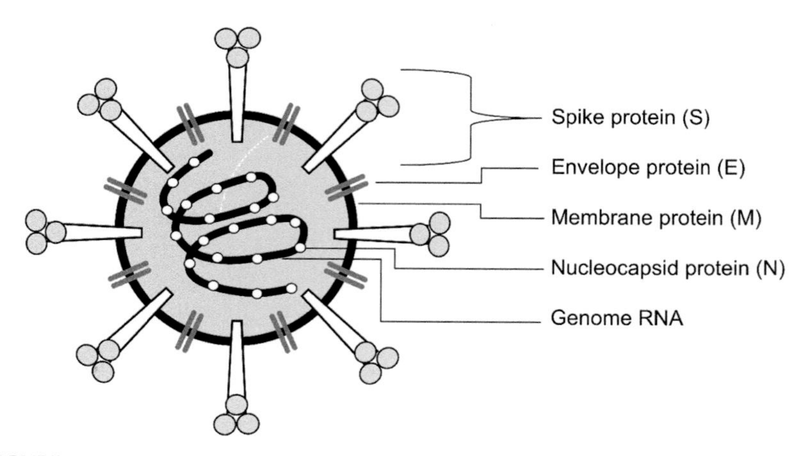

FIGURE 1.2 Structure of severe acute respiratory syndrome—associated coronavirus 2 (SARS-CoV-2) virion ((Not drawn to scale). The viral genome is encapsulated by the nucleocapsid protein (N), which is enclosed by membrane protein (M), enveloped protein (E), and the spike protein (S). The genome is one of the largest (about 30 kilobases, that is 30,000 base pairs) among all RNA viral genomes known. The outermost layer of the virus particle is made of membrane (M) and enveloped (E) protein that is derived from host cell membranes as well as virus coded.

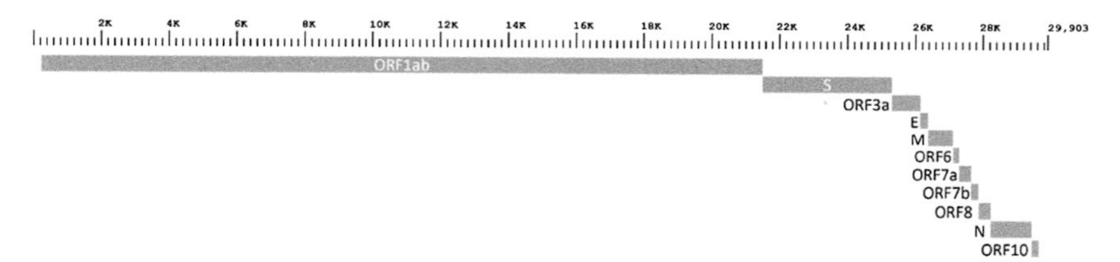

FIGURE 1.3 Genome organization of severe acute respiratory syndrome—associated coronavirus 2 (SARS-CoV-2) (open reading frame (ORF) encodes the nonstructural proteins, accounting for most of the nucleotides followed by the spike protein).

the virus to the host cell surface receptor known as the angiotensin-converting enzyme 2 (ACE2), an enzyme abundantly expressed on airway epithelial cells (type II pneumocytes and alveolar macrophages, for example [34]) and small intestinal epithelial cells) [35]. The RBD comprises the core and the receptor-binding motif (RBM). The latter is relatively less preserved than the former during viral evolution as it must evolve to sustain sufficient affinity to engage the ACE2. Biochemical data have confirmed the difference in five residues in the SARS-CoV-2 RBM, namely Y455L, L486F, N493Q, D494S, and T501N, that strengthened the binding affinity compared with that of SARS-CoV [36]. Given its role in host cell entry, the RBD is the target of 90% of the neutralizing antibody and vaccines [37]. Unfortunately, most mutations have occurred in the RBD, potentially escaping immune response and undermining vaccine efficacy. Details of SARS-CoV-2 mutations can be found in Box 1.1.

BOX 1.1 Variants of SARS-CoV-2

Like other RNA viruses, coronaviruses are more prone to error during replication that leads to mutations. Furthermore, the large genome size makes genetic recombination more frequent, further increasing genetic variability. A number of variants have emerged since the outbreak. The WHO has categorized these variants into "variants of interest" (VOI) and "variants of concern" (VOC), depending on their impact on global public health. While the former refers to the variants that have been identified to cause multiple cases/clusters or that have been detected in multiple countries, the latter signifies an increase in the transmissibility of COVID-19, a change in clinical presentations, or a decrease in effectiveness of public health measures, vaccines, or therapeutics. As of the end of 2021, five variants have been classified as variants of concern, including Beta in South Africa, Gamma in Brazil, Delta in India, and Omicron in South Africa. According to the WHO [38], the Delta strain remains the most prevalent one in most countries, accounting for at least 50% of all COVID-19 cases in Canada, the United Kingdom, the EU, Russia, India, Brazil, Argentina, Chile, South Africa, Australia, New Zealand, Peru, Ukraine, Malaysia, Indonesia, South Korea, and Japan. It contains L452R, T478K, D614G, and P681R mutations in the spike protein [39] that may affect the infectivity and resistance to antibodies. However, the recently emerged Omicron strain is much more infectious than the Beta and Delta strain [40] and has dominated other prevailing strains (Fig. 1.4). Existing evidence suggests that individuals infected with the Omicron strain are between 50% and 70% less likely to be admitted compared to the Delta strain [41]. Nevertheless, the Delta strain remains a health concern as it may reduce the efficacy of the vaccine [42].

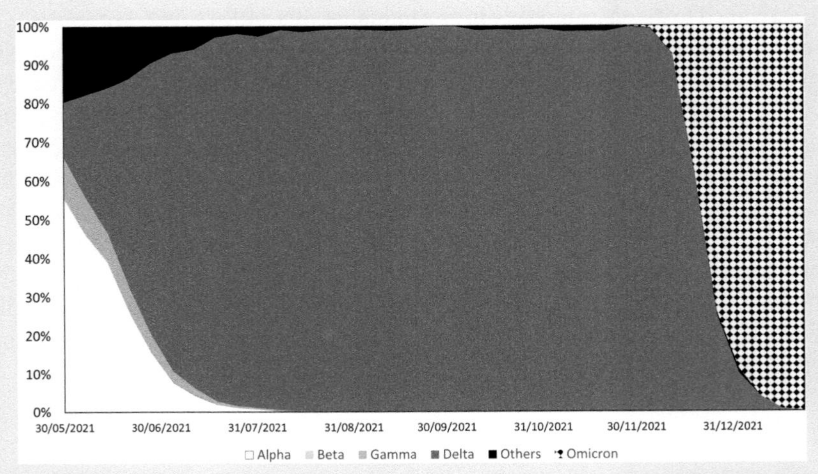

FIGURE 1.4 Prevalence of different severe acute respiratory syndrome associated coronavirus 2 (SARS-CoV-2) variants in the United States (the Omicron variant has spread much faster than the Delta variant).

Continued

Box 1.1 Variants of SARS-CoV-2—cont'd

Virologically, it has a large number of mutations, including synonymous mutations, in which the nucleotide is modified but not the amino acid, as well as nonsynonymous mutations (Fig. 1.5).

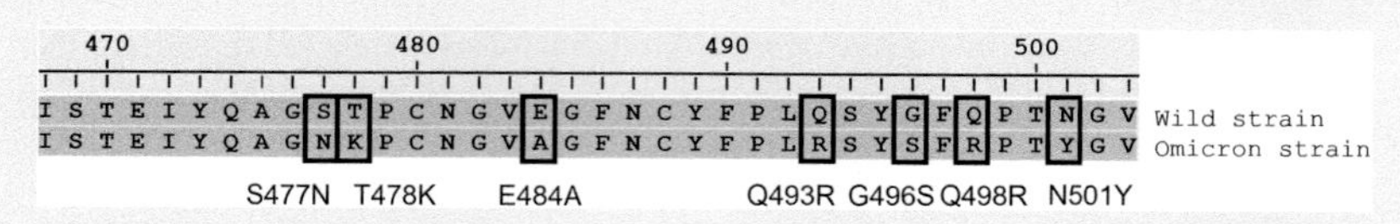

FIGURE 1.5 Sequence alignment of the wild and Omicron strain (some of the nonsynonymous mutations in the receptor-binding motif (RBM) as positions 477, 478, 484, 493, 496, 498, and 501).

In between S1 and S2 subunit is the multibasic S1/S2 site that is unique to SARS-CoV, SARS-CoV-2, and MERS-CoV and is important for virulence [43]. The S2 subunit forms the bottom part of the spike, mediating the fusion of viral and host cell membranes after priming with the transmembrane serine protease 2 (TMPRSS2) [44] expressed in type II pneumocytes [45]. After membrane fusion, the nucleocapsid protein is released into the host cell cytoplasm, and the viral RNA is uncoated for transcription.

Coupled with the postmortem examination finding of viral cytopathic effect in type II pneumocytes [46], the expression amount of ACE2 and TMPRSS2 has been linked to the susceptibility to and mortality of SARS-CoV-2. For instance, it has been suggested that the higher COVID-19-related mortality in the elderly was attributed to their higher expression of ACE2 [34]. Findings on the difference in ACE2 expression between children and adults remain contradicting and are subject to debate. A study found that children aged below 10 years had lower nasal gene expression of ACE2 compared to other age groups [47], whereas no difference was observed between children below 3 years of age and adults in another study [48]. Nasal expression of ACE2 is similar between term and preterm newborns but lower compared with adults [49]. It has been hypothesized that the finding of higher mortality risk in COVID-19 patients with Down syndrome [50] was attributed to TMPRSS2 that locates on 21q22.3, part of the Down syndrome critical region [51].

Transmission of SARS-CoV-2

Droplets, aerosols, and respiratory secretions

Severe acute respiratory syndrome—associated coronavirus 2 primarily spreads from human to human through the inhalation of respiratory secretions that encapsulate virus particles. A common form of respiratory secretion is droplets emitted during breathing, coughing, singing, sneezing, and speaking [52]. Respiratory droplets whose 95%–99% aqueous fraction evaporates can be transformed into long-lived respiratory aerosols [52], resulting in airborne transmission [53,54]. The definition of droplets and aerosols draws on the difference in their physical properties. In general, an aerosol is believed to consist of a range of small droplets (also called droplet nuclei) of size less than 5–10 μm [55,56] whose motion depends on factors such as airflow, temperature, humidity, and gravity [55]. Unlike droplets that rapidly fall to the ground after emission, an aerosol can suspend and linger in the air for a longer time and travel a longer distance. Indoor aerosol transmission has been well documented, indoor spaces with poor ventilation in particular [57]. For example, probable cases of transmission between rooms within a building have been reported in China [58], South Korea [59], and Hong Kong (Box 1.2). A social distance of 1–2 m in the absence of face masks has been recommended by the WHO [63]. However, it is important to note that sneezed droplets can travel as far as 8 m [64]. Aerosol particles can travel 9 m [65] or even 30 m if they are produced by coughing [66].

It should come to no surprise that SARS-CoV-2 RNA can be detected in the bronchoalveolar lavage fluid, sputum, saliva, nasopharyngeal, and oropharyngeal swabs of COVID-19 patients [67,68]. In particular, viral RNA is more detectable in bronchoalveolar lavage fluid than in saliva, nasopharyngeal, and oropharyngeal swabs [67–69]. Higher detectability in saliva than in oropharyngeal swabs has been observed [70], while nasopharyngeal swabs and saliva have a similar degree of detectability [71,72]. However, some studies found that saliva was less sensitive than nasopharyngeal swabs for diagnosing SARS-CoV-2 in children [73], and saliva should be used as an adjunct to oropharyngeal-nasal swabs for

BOX 1.2 Probable aerosol transmission of SARS-CoV-2

An outbreak of COVID-19 was reported in a 34-storey residential building in Richland Gardens of Hong Kong in December 2020. A total of seven individuals were tested positive for SARS-CoV-2. All patients lived in vertically aligned apartments of the building connected by drainage pipes (Fig. 1.6). Because no residents of apartments in other vertically aligned columns were tested positive despite the shared use of elevators and other communal facilities; aerosol transmission via feces was highly suspected. Of the 22 environmental samples collected in the affected apartments, eight were tested for the virus, including samples from the ventilation fans in the bathrooms [60]. Given the experience of the SARS-CoV outbreak in Amoy Gardens in Hong Kong in 2003 [61,62], engineers were consulted, and it was believed that the antisiphonage pipe connected to the drainage pipe in the bathroom was modified. As a result, air from the main drainage pipe connecting the toilets of other apartments was introduced to the bathroom, suggesting probable aerosol transmission.

Continued

Box 1.2 Probable aerosol transmission of SARS-CoV-2—cont'd

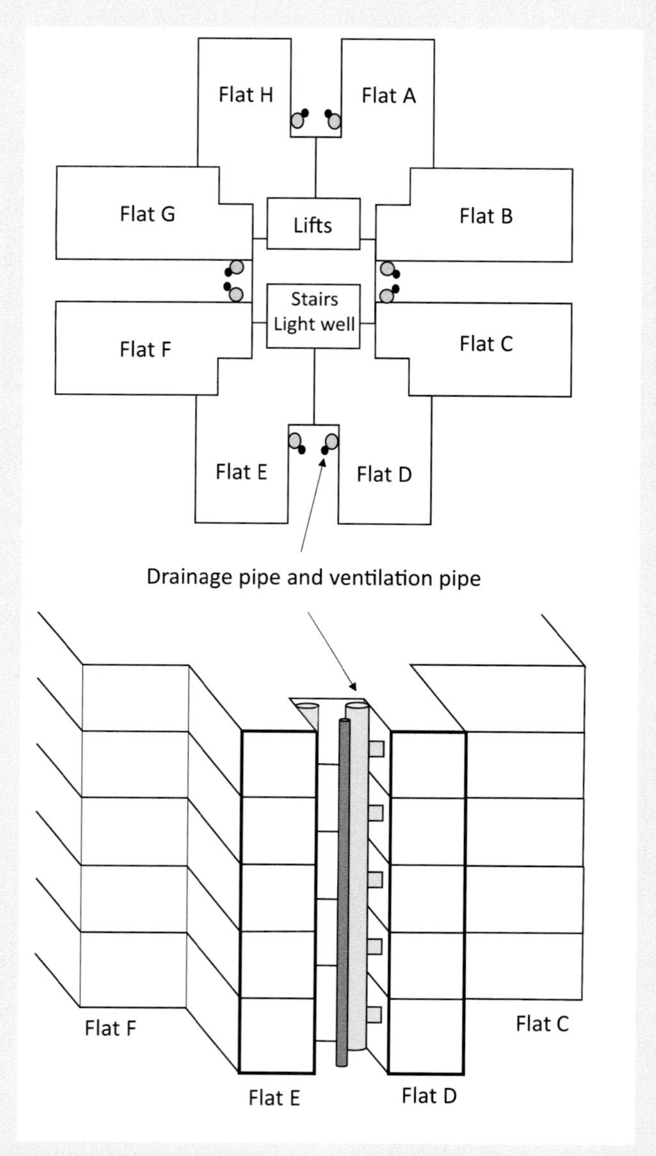

FIGURE 1.6 The floor plan and isometric view of Block 6 of Richland Gardens in Hong Kong ((*upper panel*) A typical floor of the building showing the layout of apartments, communal facilities, and drainage pipes. (*lower panel*) The main drainage pipe connects the bathrooms of vertically aligned apartments.).

testing [74]. Furthermore, rectal swabs should also be used given the extended gastrointestinal viral shedding in children (discussed in the latter section), implying that positive PCR test result on rectal swabs but negative on nasopharyngeal ones is possible [75].

Fomites

Fomite transmission is still under investigation [76]. An experimental study controlling for environmental conditions suggested that viruses in mucus and sputum can remain infectious for 24 h and detectable for a week with low temperature and humidity providing a favorable environment for the virus [77]. Surface materials also play a role in viral viability. For example, infectious virus particles can survive on copper and stainless steel for 4 and 12 h, respectively [78]. In contrast, viruses in fomite can be still be detected on the surface of cardboard and polypropylene plastic after 24 and 96 h, respectively [79,80]. Nevertheless, the lack of evidence demonstrating the recovery of viable virus in most of these studies indicates low risks of fomite transmission [81], despite transmissions and outbreaks linked to contaminated food have been suspected in the United States and Hong Kong [82].

Other types of human secretions

Although SARS-CoV-2 often infects syncytiotrophoblasts in pregnant patients as demonstrated by pathological examinations [83], existing evidence indicates low risk of vertical transmission [84], albeit with cases reported [85,86]. Some studies have reported the detection of SARS-CoV-2 in breast milk [87,88] and vaginal fluids [89] from women positive for SARS-CoV-2 RNA. There are also studies reporting the detection of SARS-CoV-2 in urine [90], semen [91,92], blood [93], tears, and cerumen [94]. Nevertheless, there is little evidence of transmission through contact with these human secretions.

It is worth noting that SARS-CoV-2 has been isolated from stool samples from COVID-19 patients [95,96]. In a multicenter study in China, SARS-CoV-2 RNA was detected in over 90% of the stool samples from pediatric patients (perhaps due to the higher intestinal expression of ACE2 in children than adults [48]), and two-thirds of the cases remained positive 14 days after discharge [97]. Based on the test results of anal and rectal swabs from 69 cases, a review suggested an average gastrointestinal tract viral shedding duration of 24 days [98]. In another review study, viral shedding was 9 days longer in fecal than in respiratory samples [99]. Taken together, these findings indicated longer fecal shedding in children compared with adults [100,101]. Although there is still no data confirming fecal—oral transmission involving pediatric patients [102], hand hygiene should still be performed regularly.

Zoonosis

There have been reports of transmission from humans to Felidae and viral replication in domestic cats [103] but not vice versa. Infections of pet dogs have also been reported where PCR test results for nasal and oral swabs from these dogs were positive for the first few days after the last exposure to their infected owners [104]. In addition, transmission from minks to humans has been suspected [105]. Nonetheless, the risk of transmission from domestic pets to humans remains very low.

Transmission among children and adolescents

Existing evidence suggests that children play a limited role in transmitting the virus in both school and household settings [101] and that there is rather stronger evidence of staff to staff transmission in schools [106]. While it is unclear why children do not appear to drive transmissions, a number of hypotheses have been proposed. First, smaller class sizes might have impacts on the spread of the virus, as indicated by the more robust transmission in high schools than in primary schools [107,108]. Second, asymptomaticitiy in children might have underestimated the role of children in the transmission of virus as they are less likely to get tested and hospitalized [101]. Finally, the higher risk of infection in households than in schools might be attributed to the adherence to preventive measures in schools, such as frequent hand hygiene, social distancing, and the use of face masks and personal protective equipment (PPEs) [109].

Other epidemiological characteristics of COVID-19

Incubation period

From a clinical perspective, the incubation period of a disease can be used to estimate the date of exposure for mandatory reporting of many infectious diseases and to assess the risk of infection when it is used as a case definition. In general, the average incubation period of COVID-19 is usually between 4 and 6 days [110,111]. This is slightly longer than the incubation period of other commonly circulated viruses such as rhinoviruses (average of 1.9 days), influenza virus (1.4 days for influenza A and 0.6 days for influenza B), and parainfluenza virus (2.6 days) [112]. It is important to note that the incubation period varies between individuals with increased uncertainty in cases of longer-than-usual incubation periods [113]. Therefore, the average of 4–6 days should only be taken as a rule-of-thumb.

Geographical differences in the incubation period have been observed, but the average generally falls within the range [114]. Significant differences in incubation period by age have been reported with longer incubation periods in the elderly [115,116]. In contrast, the difference in the incubation period between children and adults requires further investigations because of the lack of solid evidence. A few studies have reported slightly longer incubation periods in the age of 6–9 days [97,117–120]. Because presymptomatic and asymptomatic patients can be infectious [121,122], virus transmissions can occur during the incubation period. It has been suggested that the viral load peaks a day before to 7 days after the symptom onset [123], meaning that PCR test results can be positive before symptom onset [120].

Mortality and morbidity risk factors in pediatric patients

It is generally known that the underage are less susceptible to SARS-CoV-2 infection and fatality than adults. However, recent evidence has emerged that infants and newborns are more vulnerable to mortality and morbidity compared with children and adolescents. A study in multiple countries found a U-shaped relationship between COVID-19-related death and age with higher mortality rates at the beginning and old age of life [124]. Similarly, a nationwide study in Brazil has found that patients aged below 2 years and aged between 12 and 19 years have a similar risk of death and is higher than that in those aged between 2 and 11 years [125]. At least two systematic reviews have concluded that infants and neonates are more likely to have a severe clinical course compared with children [126,127]. Studies have suggested different ages below which the mortality or morbidity risk increases, including 30 days [128], 1 [129], and 2 years [125].

Research concerning the impact of comorbidities on pediatric patients remains very scant. For instance, there is limited data on the impact of pediatric asthma on SARS-CoV-2 infection [130,131]. For asthma and COVID-19 in adults, findings concerning the risk of infection [132,133] and mortality [134–137] are highly conflicting. Current evidence does not suggest any benefits or harms of inhaled corticosteroids for SARS-CoV-2 infection [134,138].

Down syndrome has been identified as a mortality and morbidity risk factor in adults with SARS-CoV-2. Based on the data of over 8 million adults in the United Kingdom, a study has estimated a 4-fold and 10-fold increase in hazard of COVID-19-related hospitalization and death in those with Down syndrome relative to those without, respectively [50]. Little has been done to understand the role of Down syndrome in pediatric patients. A study suggested that, among the 328 COVID-19 pediatric patients with Down syndrome from a variety of countries, those with obesity and epilepsy had 2.3 and 4.0 times, respectively, the odds of hospitalization compared with those without [139]. Using controls from the United States only, the authors suggested no significant difference in the prevalence of multisystem inflammatory syndrome and ICU admission rate between those with and without Down syndrome.

Based on existing studies of low-level evidence, little is known about SARS-CoV-2 infection in patients with congenital heart disease (CHD). A study summarizing a few case reports and case series [140] concluded that pediatric patients with CHD may be more prone to a more severe clinical course of COVID-19, and that Down syndrome with common atrioventricular canal is the most common form of CHD in these patients.

Obesity is a risk factor for COVID-19-related mortality in the general population [141]. Similarly, it has been suggested that obesity is associated with an increased risk of prolonged respiratory symptoms and mortality in adolescents with SARS-CoV-2, compared with those without obesity [142, 143].

Conclusion

This chapter reviews the medical virological aspects of coronaviruses pathogenic to humans, including the causative agent of common colds and ARDS, with the emphasis on the structure and host cell entry of SARS-CoV-2. Modes of the transmission of SARS-CoV-2 are intensively discussed. Respiratory secretions remain the primary means of virus transmission despite positive detection of viral RNA in specimens of other types of human secretions. Epidemiological characteristics of SARS-CoV-2 infection in the underage still require further investigation. Some studies have observed slightly longer incubation periods in children. While more recent studies have reported an increased risk of mortality in newborns and infants, little is known about the impact of comorbidities on pediatric patients, asthma, and Down syndrome, for example.

References

[1] Huang C, Wang Y, Li X, Ren L, Zhao J, Hu Y, Zhang L, Fan G, Xu J, Gu X, Cheng Z, Yu T, Xia J, Wei Y, Wu W, Xie X, Yin W, Li H, Liu M, Cao B. Clinical features of patients infected with 2019 novel coronavirus in Wuhan, China. Lancet (London, England) 2020a;395(10223):497–506. https://doi.org/10.1016/S0140-6736(20)30183-5.

[2] Zanin M, Xiao C, Liang T, Ling S, Zhao F, Huang Z, Lin F, Lin X, Jiang Z, Wong S-S. The public health response to the COVID-19 outbreak in mainland China: a narrative review. J Thorac Dis 2020;12(8):4434–49. https://doi.org/10.21037/jtd-20-2363.

[3] China CDC detects new coronaviruses at the Huanan Seafood Wholesale Market in Wuhan. [Media release]. Chinese Center for Disease Control and Prevention. 2020. https://www.chinacdc.cn/yw_9324/202001/t20200127_211469.html (Original work published 2020).

[4] Ji W, Wang W, Zhao X, Zai J, Li X. Cross-species transmission of the newly identified coronavirus 2019-nCoV. J Med Virol 2020;92(4):433–40. https://doi.org/10.1002/jmv.25682.

[5] Bukhari K, Mulley G, Gulyaeva AA, Zhao L, Shu G, Jiang J, Neuman BW. Description and initial characterization of metatranscriptomic nidovirus-like genomes from the proposed new family Abyssoviridae, and from a sister group to the Coronavirinae, the proposed genus Alphaletovirus. Virology 2018;524:160–71. https://doi.org/10.1016/j.virol.2018.08.010.

[6] Helmy YA, Fawzy M, Elaswad A, Sobieh A, Kenney SP, Shehata AA. The COVID-19 pandemic: a comprehensive review of taxonomy, genetics, epidemiology, diagnosis, treatment, and control. J Clin Med 2020;9(4). https://doi.org/10.3390/jcm9041225.

[7] Wang W, Lin X-D, Liao Y, Guan X-Q, Guo W-P, Xing J-G, Holmes EC, Zhang Y-Z. Discovery of a highly divergent coronavirus in the asian house shrew from China illuminates the origin of the alphacoronaviruses. J Virol 2017;91(17). https://doi.org/10.1128/JVI.00764-17.

[8] Zhou Z, Qiu Y, Ge X. Order. Animal Diseases 2021;1(1):5. https://doi.org/10.1186/s44149-021-00005-9.

[9] Tyrrell DA, Bynoe ML. Cultivation of viruses from a high proportion of patients with colds. Lancet (London, England) 1966;1(7428):76–7.

[10] Hamre D, Procknow JJ. A new virus isolated from the human respiratory tract. Proc Soc Exp Biol Med 1966;121(1):190–3.

[11] van der Hoek L, Pyrc K, Jebbink MF, Vermeulen-Oost W, Berkhout RJM, Wolthers KC, Wertheim-van Dillen PME, Kaandorp J, Spaargaren J, Berkhout B. Identification of a new human coronavirus. Nat Med 2004;10(4):368–73.

[12] Woo PCY, Lau SKP, Chu C, Chan K, Tsoi H, Huang Y, Wong BHL, Poon RWS, Cai JJ, Luk W, Poon LLM, Wong SSY, Guan Y, Peiris JSM, Yuen K. Characterization and complete genome sequence of a novel coronavirus, coronavirus HKU1, from patients with pneumonia. J Virol 2005;79(2):884–95.

[13] Corman VM, Muth D, Niemeyer D, Drosten C. Hosts and sources of endemic human coronaviruses. Adv Virus Res 2018;100:163–88. https://doi.org/10.1016/bs.aivir.2018.01.001.

[14] Eccles R. Understanding the symptoms of the common cold and influenza. Lancet Infect Dis 2005;5(11):718–25.

[15] Heimdal I, Moe N, Krokstad S, Christensen A, Skanke LH, Nordbø SA, Døllner H. Human coronavirus in hospitalized children with respiratory tract infections: a 9-year population-based study from Norway. J Infect Dis 2019;219(8):1198–206. https://doi.org/10.1093/infdis/jiy646.

[16] Chiu SS, Chan KH, Chu KW, Kwan SW, Guan Y, Poon LLM, Peiris JSM. Human coronavirus NL63 infection and other coronavirus infections in children hospitalized with acute respiratory disease in Hong Kong, China. Clin Infect Dis 2005;40(12):1721–9.

[17] Kuypers J, Martin ET, Heugel J, Wright N, Morrow R, Englund JA. Clinical disease in children associated with newly described coronavirus subtypes. Pediatrics 2007;119(1):e70–6.

[18] Zeng Z-Q, Chen D-H, Tan W-P, Qiu S-Y, Xu D, Liang H-X, Chen M-X, Li X, Lin Z-S, Liu W-K, Zhou R. Epidemiology and clinical characteristics of human coronaviruses OC43, 229E, NL63, and HKU1: a study of hospitalized children with acute respiratory tract infection in Guangzhou, China. Eur J Clin Microbiol Infect Dis 2018;37(2):363–9. https://doi.org/10.1007/s10096-017-3144-z.

[19] Chan CM, Tse H, Wong SSY, Woo PCY, Lau SKP, Chen L, Zheng BJ, Huang JD, Yuen KY. Examination of seroprevalence of coronavirus HKU1 infection with S protein-based ELISA and neutralization assay against viral spike pseudotyped virus. J Clin Virol 2009;45(1):54–60. https://doi.org/10.1016/j.jcv.2009.02.011.

[20] Dijkman R, Jebbink MF, El Idrissi NB, Pyrc K, Müller MA, Kuijpers TW, Zaaijer HL, van der Hoek L. Human coronavirus NL63 and 229E seroconversion in children. J Clin Microbiol 2008;46(7):2368–73. https://doi.org/10.1128/JCM.00533-08.

[21] Dijkman R, Jebbink MF, Gaunt E, Rossen JWA, Templeton KE, Kuijpers TW, van der Hoek L. The dominance of human coronavirus OC43 and NL63 infections in infants. J Clin Virol 2012;53(2):135–9. https://doi.org/10.1016/j.jcv.2011.11.011.

[22] Hovi T, Kainulainen H, Ziola B, Salmi A. OC43 strain-related coronavirus antibodies in different age groups. J Med Virol 1979;3(4):313–20.

[23] Tamminen K, Salminen M, Blazevic V. Seroprevalence and SARS-CoV-2 cross-reactivity of endemic coronavirus OC43 and 229E antibodies in Finnish children and adults. Clin Immunol 2021;229:108782. https://doi.org/10.1016/j.clim.2021.108782.

[24] Zhao G. SARS molecular epidemiology: a Chinese fairy tale of controlling an emerging zoonotic disease in the genomics era. Philos Trans R Soc Lon B Biol Sci 2007;362(1482):1063−81.

[25] SARS virus close to conclusive identification, new tests for rapid diagnosis ready soon. 2003 [Media release]. World Health Organization, https://www.who.int/emergencies/disease-outbreak-news/item/2003_03_27b-en (Original work published 2003).

[26] Donnelly CA, Ghani AC, Leung GM, Hedley AJ, Fraser C, Riley S, Abu-Raddad LJ, Ho L-M, Thach T-Q, Chau P, Chan K-P, Lam T-H, Tse L-Y, Tsang T, Liu S-H, Kong JHB, Lau EMC, Ferguson NM, Anderson RM. Epidemiological determinants of spread of causal agent of severe acute respiratory syndrome in Hong Kong. Lancet (London, England) 2003;361(9371):1761−6.

[27] Principi N, Bosis S, Esposito S. Effects of coronavirus infections in children. Emerg Infect Dis 2010;16(2):183−8. https://doi.org/10.3201/eid1602.090469.

[28] Al-Abdallat MM, Payne DC, Alqasrawi S, Rha B, Tohme RA, Abedi GR, Al Nsour M, Iblan I, Jarour N, Farag NH, Haddadin A, Al-Sanouri T, Tamin A, Harcourt JL, Kuhar DT, Swerdlow DL, Erdman DD, Pallansch MA, Haynes LM, Gerber SI. Hospital-associated outbreak of Middle East respiratory syndrome coronavirus: a serologic, epidemiologic, and clinical description. Clin Infect Dis 2014;59(9):1225−33. https://doi.org/10.1093/cid/ciu359.

[29] Zaki AM, van Boheemen S, Bestebroer TM, Osterhaus ADME, Fouchier RAM. Isolation of a novel coronavirus from a man with pneumonia in Saudi Arabia. N Engl J Med 2012;367(19):1814−20. https://doi.org/10.1056/NEJMoa1211721.

[30] Zhang A-R, Shi W-Q, Liu K, Li X-L, Liu M-J, Zhang W-H, Zhao G-P, Chen J-J, Zhang X-A, Miao D, Ma W, Liu W, Yang Y, Fang L-Q. Epidemiology and evolution of Middle East respiratory syndrome coronavirus, 2012-2020. Infect Dis Poverty 2021;10(1):66. https://doi.org/10.1186/s40249-021-00853-0.

[31] Thabet F, Chehab M, Bafaqih H, Al Mohaimeed S. Middle East respiratory syndrome coronavirus in children. Saudi Med J 2015;36(4):484−6. https://doi.org/10.15537/smj.2015.4.10243.

[32] MacIntyre CR, Chen X, Adam DC, Chughtai AA. Epidemiology of paediatric Middle East respiratory syndrome coronavirus and implications for the control of coronavirus virus disease 2019. J Paediatr Child Health 2020;56(10):1561−4. https://doi.org/10.1111/jpc.15014.

[33] Zhu N, Zhang D, Wang W, Li X, Yang B, Song J, Zhao X, Huang B, Shi W, Lu R, Niu P, Zhan F, Ma X, Wang D, Xu W, Wu G, Gao GF, Tan W. A novel coronavirus from patients with pneumonia in China, 2019. N Engl J Med 2020a;382(8):727−33. https://doi.org/10.1056/NEJMoa2001017.

[34] Baker SA, Kwok S, Berry GJ, Montine TJ. Angiotensin-converting enzyme 2 (ACE2) expression increases with age in patients requiring mechanical ventilation. PLoS One 2021;16(2):e0247060. https://doi.org/10.1371/journal.pone.0247060.

[35] Bourgonje AR, Abdulle AE, Timens W, Hillebrands J-L, Navis GJ, Gordijn SJ, Bolling MC, Dijkstra G, Voors AA, Osterhaus AD, van der Voort PH, Mulder DJ, van Goor H. Angiotensin-converting enzyme 2 (ACE2), SARS-CoV-2 and the pathophysiology of coronavirus disease 2019 (COVID-19). J Pathol 2020;251(3):228−48. https://doi.org/10.1002/path.5471.

[36] Hu B, Guo H, Zhou P, Shi Z-L. Characteristics of SARS-CoV-2 and COVID-19. Nat Rev Microbiol 2021;19(3):141−54. https://doi.org/10.1038/s41579-020-00459-7.

[37] Piccoli L, Park Y-J, Tortorici MA, Czudnochowski N, Walls AC, Beltramello M, Silacci-Fregni C, Pinto D, Rosen LE, Bowen JE, Acton OJ, Jaconi S, Guarino B, Minola A, Zatta F, Sprugasci N, Bassi J, Peter A, De Marco A, Veesler D. Mapping neutralizing and immunodominant sites on the SARS-CoV-2 spike receptor-binding domain by structure-guided high-resolution serology. Cell 2020;183(4):1024–1042.e21. https://doi.org/10.1016/j.cell.2020.09.037.

[38] COVID-19 Weekly Epidemiological Update. [Situation report]. World Health Organization. 2021. https://www.who.int/docs/default-source/coronaviruse/situation-reports/20211109_weekly_epi_update_65.pdf?sfvrsn=1fa9c39c_3&download=true (Original work published 2021).

[39] Bian L, Gao Q, Gao F, Wang Q, He Q, Wu X, Mao Q, Xu M, Liang Z. Impact of the Delta variant on vaccine efficacy and response strategies. Expet Rev Vaccine 2021;20(10):1201–9. https://doi.org/10.1080/14760584.2021.1976153.

[40] He X, Hong W, Pan X, Lu G, Wei X. SARS-CoV-2 Omicron variant: Characteristics and prevention. MedComm (2020) 2021;2(4):838–45. https://doi.org/10.1002/mco2.110.

[41] Mahase E. Covid-19: hospital admission 50–70% less likely with omicron than delta, but transmission a major concern. BMJ (Clin Res Ed) 2021;375:n3151. https://doi.org/10.1136/bmj.n3151.

[42] Christie B. Covid-19: early studies give hope omicron is milder than other variants. BMJ (Clin Res Ed) 2021;375:n3144. https://doi.org/10.1136/bmj.n3144.

[43] Hoffmann M, Kleine-Weber H, Pöhlmann S. A multibasic cleavage site in the spike protein of SARS-CoV-2 is essential for infection of human lung cells. Mol Cell 2020;78(4):779–784.e5. https://doi.org/10.1016/j.molcel.2020.04.022.

[44] Duan L, Zheng Q, Zhang H, Niu Y, Lou Y, Wang H. The SARS-CoV-2 spike glycoprotein biosynthesis, structure, function, and antigenicity: implications for the design of spike-based vaccine immunogens. Front Immunol 2020;11:576622. https://doi.org/10.3389/fimmu.2020.576622.

[45] Glowacka I, Bertram S, Müller MA, Allen P, Soilleux E, Pfefferle S, Steffen I, Tsegaye TS, He Y, Gnirss K, Niemeyer D, Schneider H, Drosten C, Pöhlmann S. Evidence that TMPRSS2 activates the severe acute respiratory syndrome coronavirus spike protein for membrane fusion and reduces viral control by the humoral immune response. J Virol 2011;85(9):4122–34. https://doi.org/10.1128/JVI.02232-10.

[46] Oprinca G-C, Muja L-A. Postmortem examination of three SARS-CoV-2-positive autopsies including histopathologic and immunohistochemical analysis. Int J Leg Med 2021;135(1):329–39. https://doi.org/10.1007/s00414-020-02406-w.

[47] Bunyavanich S, Do A, Vicencio A. Nasal gene expression of angiotensin-converting enzyme 2 in children and adults. JAMA 2020;323(23):2427–9. https://doi.org/10.1001/jama.2020.8707.

[48] Berni Canani R, Comegna M, Paparo L, Cernera G, Bruno C, Strisciuglio C, Zollo I, Gravina AG, Miele E, Cantone E, Gennarelli N, Nocerino R, Carucci L, Giglio V, Amato F, Castaldo G. Age-related differences in the expression of most relevant mediators of SARS-CoV-2 infection in human respiratory and gastrointestinal tract. Front Pediatr 2021;9:697390. https://doi.org/10.3389/fped.2021.697390.

[49] Heinonen S, Helve O, Andersson S, Janér C, Süvari L, Kaskinen A. Nasal expression of SARS-CoV-2 entry receptors in newborns. Arch Dis Child Fetal Neonatal Ed 2021;107(1):95–7. https://doi.org/10.1136/archdischild-2020-321334.

[50] Clift AK, Coupland CAC, Keogh RH, Hemingway H, Hippisley-Cox J. COVID-19 mortality risk in Down syndrome: results from a cohort study of 8 million adults. Ann Intern Med 2021;174(4):572–6. https://doi.org/10.7326/M20-4986.

[51] Hou Y, Zhao J, Martin W, Kallianpur A, Chung MK, Jehi L, Sharifi N, Erzurum S, Eng C, Cheng F. New insights into genetic susceptibility of COVID-19: an ACE2 and TMPRSS2 polymorphism analysis. BMC Med 2020;18(1):216. https://doi.org/10.1186/s12916-020-01673-z.

[52] Stadnytskyi V, Anfinrud P, Bax A. Breathing, speaking, coughing or sneezing: what drives transmission of SARS-CoV-2? J Intern Med 2021;290(5):1010−27. https://doi.org/10.1111/joim.13326.

[53] Delikhoon M, Guzman MI, Nabizadeh R, Norouzian Baghani A. Modes of transmission of severe acute respiratory syndrome-coronavirus-2 (SARS-CoV-2) and factors influencing on the airborne transmission: a review. Int J Environ Res Publ Health 2021;18(2). https://doi.org/10.3390/ijerph18020395.

[54] Tang S, Mao Y, Jones RM, Tan Q, Ji JS, Li N, Shen J, Lv Y, Pan L, Ding P, Wang X, Wang Y, MacIntyre CR, Shi X. Aerosol transmission of SARS-CoV-2? Evidence, prevention and control. Environ Int 2020;144:106039. https://doi.org/10.1016/j.envint.2020.106039.

[55] Atkinson J, Chartier Y, Pessoa-Silva CL, Jensen P, Li Y, Seto W-H. Annex C. Respiratory droplets. In: Natural ventilation for infection control in health-care settings. World Health Organization; 2009. https://www.who.int/water_sanitation_health/publications/natural_ventilation.pdf.

[56] Smieszek T, Lazzari G, Salathé M. Assessing the dynamics and control of droplet- and aerosol-transmitted influenza using an indoor positioning system. Sci Rep 2019;9(1):2185. https://doi.org/10.1038/s41598-019-38825-y.

[57] Bazant MZ, Bush JWM. A guideline to limit indoor airborne transmission of COVID-19. Proc Natl Acad Sci USA 2021;118(17). https://doi.org/10.1073/pnas.2018995118.

[58] Kang M, Wei J, Yuan J, Guo J, Zhang Y, Hang J, Qu Y, Qian H, Zhuang Y, Chen X, Peng X, Shi T, Wang J, Wu J, Song T, He J, Li Y, Zhong N. Probable evidence of fecal aerosol transmission of SARS-CoV-2 in a high-rise building. Ann Intern Med 2020;173(12):974−80. https://doi.org/10.7326/M20-0928.

[59] Hwang SE, Chang JH, Oh B, Heo J. Possible aerosol transmission of COVID-19 associated with an outbreak in an apartment in Seoul, South Korea, 2020. Int J Infect Dis 2021;104:73−6. https://doi.org/10.1016/j.ijid.2020.12.035.

[60] 8 unit samples test positive. Press conference. The Government of Hong Kong. 2020. https://www.news.gov.hk/eng/2020/12/20201211/20201211_171140_100.html.

[61] McKinney KR, Gong YY, Lewis TG. Environmental transmission of SARS at Amoy Gardens. J Environ Health 2006;68(9):26−30. quiz 51−52.

[62] Yu ITS, Li Y, Wong TW, Tam W, Chan AT, Lee JHW, Leung DYC, Ho T. Evidence of airborne transmission of the severe acute respiratory syndrome virus. N Engl J Med 2004;350(17):1731−9.

[63] Setti L, Passarini F, De Gennaro G, Barbieri P, Perrone MG, Borelli M, Palmisani J, Di Gilio A, Piscitelli P, Miani A. Airborne transmission route of COVID-19: why 2 meters/6 feet of inter-personal distance could not Be enough. Int J Environ Res Publ Health 2020;17(8). https://doi.org/10.3390/ijerph17082932.

[64] Zhou M, Zou J. A dynamical overview of droplets in the transmission of respiratory infectious diseases. Phys Fluids (Woodbury, N.Y.: 1994) 2021;33(3):031301. https://doi.org/10.1063/5.0039487.

[65] Ranga U. SARS-CoV-2 aerosol and droplets: an overview. Virusdisease 2021:1−8. https://doi.org/10.1007/s13337-021-00660-z.

[66] Gorbunov B. Aerosol particles generated by coughing and sneezing of a SARS-CoV-2 (COVID-19) host travel over 30 m distance. Aerosol Air Qual Res 2021;21(3):200468. https://doi.org/10.4209/aaqr.200468.

[67] Khiabani K, Amirzade-Iranaq MH. Are saliva and deep throat sputum as reliable as common respiratory specimens for SARS-CoV-2 detection? A systematic review and meta-analysis. Am J Infect Control 2021;49(9):1165–76. https://doi.org/10.1016/j.ajic.2021.03.008.

[68] Mohammadi A, Esmaeilzadeh E, Li Y, Bosch RJ, Li JZ. SARS-CoV-2 detection in different respiratory sites: a systematic review and meta-analysis. EBioMedicine 2020;59:102903. https://doi.org/10.1016/j.ebiom.2020.102903.

[69] Dentone C, Vena A, Loconte M, Grillo F, Brunetti I, Barisione E, Tedone E, Mora S, Di Biagio A, Orsi A, De Maria A, Nicolini L, Ball L, Giacobbe DR, Magnasco L, Delfino E, Mastracci L, Mangerini R, Taramasso L, Bassetti M. Bronchoalveolar lavage fluid characteristics and outcomes of invasively mechanically ventilated patients with COVID-19 pneumonia in Genoa, Italy. BMC Infect Dis 2021;21(1):353. https://doi.org/10.1186/s12879-021-06015-9.

[70] Lee RA, Herigon JC, Benedetti A, Pollock NR, Denkinger CM. Performance of saliva, oropharyngeal swabs, and nasal swabs for SARS-CoV-2 molecular detection: a systematic review and meta-analysis. J Clin Microbiol 2021;59(5). https://doi.org/10.1128/JCM.02881-20.

[71] Butler-Laporte G, Lawandi A, Schiller I, Yao M, Dendukuri N, McDonald EG, Lee TC. Comparison of saliva and nasopharyngeal swab nucleic acid amplification testing for detection of SARS-CoV-2: a systematic review and meta-analysis. JAMA Intern Med 2021;181(3):353–60. https://doi.org/10.1001/jamainternmed.2020.8876.

[72] Fan G, Qin X, Streblow DN, Hoyos CM, Hansel DE. Comparison of SARS-CoV-2 PCR-based detection using saliva or nasopharyngeal swab specimens in asymptomatic populations. Microbiol Spectr 2021;9(1):e0006221. https://doi.org/10.1128/Spectrum.00062-21.

[73] von Linstow M-L, Kruse A, Kirkby N, Marie Søes L, Nygaard U, Poulsen A. Saliva is inferior to nose and throat swabs for SARS-CoV-2 detection in children. Acta Paediatr 2021;110(12):3325–6. https://doi.org/10.1111/apa.16049.

[74] Oliver J, Tosif S, Lee L-Y, Costa A-M, Bartel C, Last K, Clifford V, Daley A, Allard N, Orr C, Nind A, Alexander K, Meagher N, Sait M, Ballard SA, Williams E, Bond K, Williamson DA, Crawford NW, Gibney KB. Adding saliva testing to oropharyngeal and deep nasal swab testing increases PCR detection of SARS-CoV-2 in primary care and children. Med J Aust 2021;215(6):273–8. https://doi.org/10.5694/mja2.51188.

[75] Xu Y, Li X, Zhu B, Liang H, Fang C, Gong Y, Guo Q, Sun X, Zhao D, Shen J, Zhang H, Liu H, Xia H, Tang J, Zhang K, Gong S. Characteristics of pediatric SARS-CoV-2 infection and potential evidence for persistent fecal viral shedding. Nat Med 2020b;26(4):502–5. https://doi.org/10.1038/s41591-020-0817-4.

[76] Harrison AG, Lin T, Wang P. Mechanisms of SARS-CoV-2 transmission and pathogenesis. Trends Immunol 2020;41(12):1100–15. https://doi.org/10.1016/j.it.2020.10.004.

[77] Matson MJ, Yinda CK, Seifert SN, Bushmaker T, Fischer RJ, van Doremalen N, Lloyd-Smith JO, Munster VJ. Effect of environmental conditions on SARS-CoV-2 stability in human nasal mucus and sputum. Emerg Infect Dis 2020;26(9). https://doi.org/10.3201/eid2609.202267.

[78] Ronca SE, Sturdivant RX, Barr KL, Harris D. SARS-CoV-2 viability on 16 common indoor surface finish materials. HERD 2021;14(3):49–64. https://doi.org/10.1177/1937586721991535.

[79] Pastorino B, Touret F, Gilles M, de Lamballerie X, Charrel RN. Prolonged infectivity of SARS-CoV-2 in fomites. Emerg Infect Dis 2020;26(9). https://doi.org/10.3201/eid2609.201788.
[80] van Doremalen N, Bushmaker T, Morris DH, Holbrook MG, Gamble A, Williamson BN, Tamin A, Harcourt JL, Thornburg NJ, Gerber SI, Lloyd-Smith JO, de Wit E, Munster VJ. Aerosol and surface stability of SARS-CoV-2 as compared with SARS-CoV-1. N Engl J Med 2020;382(16):1564–7. https://doi.org/10.1056/NEJMc2004973.
[81] Onakpoya IJ, Heneghan CJ, Spencer EA, Brassey J, Plüddemann A, Evans DH, Conly JM, Jefferson T. SARS-CoV-2 and the role of fomite transmission: a systematic review. F1000Research 2021;10:233. https://doi.org/10.12688/f1000research.51590.3.
[82] Hon KL, Leung KKY, Tang JW, Leung AKC, Li Y. COVID-19 in Hong Kong—public health, food safety, and animal vectors perspectives. J Virol Methods 2021;290:114036. https://doi.org/10.1016/j.jviromet.2020.114036.
[83] Komine-Aizawa S, Takada K, Hayakawa S. Placental barrier against COVID-19. Placenta 2020;99:45–9. https://doi.org/10.1016/j.placenta.2020.07.022.
[84] Bwire GM, Njiro BJ, Mwakawanga DL, Sabas D, Sunguya BF. Possible vertical transmission and antibodies against SARS-CoV-2 among infants born to mothers with COVID-19: a living systematic review. J Med Virol 2021;93(3):1361–9. https://doi.org/10.1002/jmv.26622.
[85] Deniz M, Tezer H. Vertical transmission of SARS CoV-2: a systematic review. J Matern Fetal Neonatal Med 2020:1–8. https://doi.org/10.1080/14767058.2020.1793322.
[86] Pietrasanta C, Ronchi A, Schena F, Ballerini C, Testa L, Artieri G, Mercadante D, Mosca F, Pugni L. SARS-CoV-2 infection and neonates: a review of evidence and unresolved questions. Pediatr Allergy Immunol 2020;31(Suppl. 26):79–81. https://doi.org/10.1111/pai.13349.
[87] Centeno-Tablante E, Medina-Rivera M, Finkelstein JL, Rayco-Solon P, Garcia-Casal MN, Rogers L, Ghezzi-Kopel K, Ridwan P, Peña-Rosas JP, Mehta S. Transmission of SARS-CoV-2 through breast milk and breastfeeding: a living systematic review. Ann N Y Acad Sci 2021;1484(1):32–54. https://doi.org/10.1111/nyas.14477.
[88] Chambers C, Krogstad P, Bertrand K, Contreras D, Tobin NH, Bode L, Aldrovandi G. Evaluation for SARS-CoV-2 in breast milk from 18 infected women. JAMA 2020;324(13):1347–8. https://doi.org/10.1001/jama.2020.15580.
[89] Schwartz A, Yogev Y, Zilberman A, Alpern S, Many A, Yousovich R, Gamzu R. Detection of severe acute respiratory syndrome coronavirus 2 (SARS-CoV-2) in vaginal swabs of women with acute SARS-CoV-2 infection: a prospective study. BJOG An Int J Obstet Gynaecol 2021;128(1):97–100. https://doi.org/10.1111/1471-0528.16556.
[90] Sun J, Zhu A, Li H, Zheng K, Zhuang Z, Chen Z, Shi Y, Zhang Z, Chen S-B, Liu X, Dai J, Li X, Huang S, Huang X, Luo L, Wen L, Zhuo J, Li Y, Wang Y, Li Y-M. Isolation of infectious SARS-CoV-2 from urine of a COVID-19 patient. Emerg Microb Infect 2020;9(1):991–3. https://doi.org/10.1080/22221751.2020.1760144.
[91] Li D, Jin M, Bao P, Zhao W, Zhang S. Clinical characteristics and results of semen tests among men with coronavirus disease 2019. JAMA Netw Open 2020a;3(5):e208292. https://doi.org/10.1001/jamanetworkopen.2020.8292.
[92] Massarotti C, Garolla A, Maccarini E, Scaruffi P, Stigliani S, Anserini P, Foresta C. SARS-CoV-2 in the semen: where does it come from? Andrology 2021;9(1):39–41. https://doi.org/10.1111/andr.12839.

[93] Wang W, Xu Y, Gao R, Lu R, Han K, Wu G, Tan W. Detection of SARS-CoV-2 in different types of clinical specimens. JAMA 2020;323(18):1843–4. https://doi.org/10.1001/jama.2020.3786.

[94] Hanege FM, Kocoglu E, Kalcioglu MT, Celik S, Cag Y, Esen F, Bayindir E, Pence S, Alp Mese E, Agalar C. SARS-CoV-2 presence in the saliva, tears, and cerumen of COVID-19 patients. Laryngoscope 2021;131(5):E1677–82. https://doi.org/10.1002/lary.29218.

[95] Amirian ES. Potential fecal transmission of SARS-CoV-2: current evidence and implications for public health. Int J Infect Dis : 2020;95:363–70. https://doi.org/10.1016/j.ijid.2020.04.057.

[96] Cheung KS, Hung IFN, Chan PPY, Lung KC, Tso E, Liu R, Ng YY, Chu MY, Chung TWH, Tam AR, Yip CCY, Leung K-H, Fung AY-F, Zhang RR, Lin Y, Cheng HM, Zhang AJX, To KKW, Chan K-H, Leung WK. Gastrointestinal manifestations of SARS-CoV-2 infection and virus load in fecal samples from a Hong Kong cohort: systematic review and meta-analysis. Gastroenterology 2020;159(1):81–95. https://doi.org/10.1053/j.gastro.2020.03.065.

[97] Hua C-Z, Miao Z-P, Zheng J-S, Huang Q, Sun Q-F, Lu H-P, Su F-F, Wang W-H, Huang L-P, Chen D-Q, Xu Z-W, Ji L-D, Zhang H-P, Yang X-W, Li M-H, Mao Y-Y, Ying M-Z, Ye S, Shu Q, Fu J-F. Epidemiological features and viral shedding in children with SARS-CoV-2 infection. J Med Virol 2020;92(11):2804–12. https://doi.org/10.1002/jmv.26180.

[98] Xu CLH, Raval M, Schnall JA, Kwong JC, Holmes NE. Duration of respiratory and gastrointestinal viral shedding in children with SARS-CoV-2: a systematic review and synthesis of data. Pediatr Infect Dis J 2020a;39(9):e249–56. https://doi.org/10.1097/INF.0000000000002814.

[99] Santos VS, Gurgel RQ, Cuevas LE, Martins-Filho PR. Prolonged fecal shedding of SARS-CoV-2 in pediatric patients: a quantitative evidence synthesis. J Pediatr Gastroenterol Nutr 2020;71(2):150–2. https://doi.org/10.1097/MPG.0000000000002798.

[100] Jiehao C, Jin X, Daojiong L, Zhi Y, Lei X, Zhenghai Q, Yuehua Z, Hua Z, Ran J, Pengcheng L, Xiangshi W, Yanling G, Aimei X, He T, Hailing C, Chuning W, Jingjing L, Jianshe W, Mei Z. A case series of children with 2019 novel coronavirus infection: clinical and epidemiological features. Clin Infect Dis 2020;71(6):1547–51. https://doi.org/10.1093/cid/ciaa198.

[101] Li X, Xu W, Dozier M, He Y, Kirolos A, Theodoratou E. The role of children in transmission of SARS-CoV-2: a rapid review. J Glob Health 2020b;10(1):011101. https://doi.org/10.7189/jogh.10.011101.

[102] Williams PCM, Howard-Jones AR, Hsu P, Palasanthiran P, Gray PE, McMullan BJ, Britton PN, Bartlett AW. SARS-CoV-2 in children: spectrum of disease, transmission and immunopathological underpinnings. Pathology 2020;52(7):801–8. https://doi.org/10.1016/j.pathol.2020.08.001.

[103] Halfmann PJ, Hatta M, Chiba S, Maemura T, Fan S, Takeda M, Kinoshita N, Hattori S-I, Sakai-Tagawa Y, Iwatsuki-Horimoto K, Imai M, Kawaoka Y. Transmission of SARS-CoV-2 in domestic cats. N Engl J Med 2020;383(6):592–4. https://doi.org/10.1056/NEJMc2013400.

[104] Sit THC, Brackman CJ, Ip SM, Tam KWS, Law PYT, To EMW, Yu VYT, Sims LD, Tsang DNC, Chu DKW, Perera RAPM, Poon LLM, Peiris M. Infection of dogs with SARS-CoV-2. Nature 2020;586(7831):776–8. https://doi.org/10.1038/s41586-020-2334-5.

[105] Hammer AS, Quaade ML, Rasmussen TB, Fonager J, Rasmussen M, Mundbjerg K, Lohse L, Strandbygaard B, Jørgensen CS, Alfaro-Núñez A, Rosenstierne MW, Boklund A, Halasa T, Fomsgaard A, Belsham GJ, Bøtner A. SARS-CoV-2 transmission between mink

(Neovison vison) and humans, Denmark. Emerg Infect Dis 2021;27(2):547−51. https://doi.org/10.3201/eid2702.203794.

[106] Ismail SA, Saliba V, Lopez Bernal J, Ramsay ME, Ladhani SN. SARS-CoV-2 infection and transmission in educational settings: a prospective, cross-sectional analysis of infection clusters and outbreaks in England. Lancet Infect Dis 2021;21(3):344−53. https://doi.org/10.1016/S1473-3099(20)30882-3.

[107] Flasche S, Edmunds WJ. The role of schools and school-aged children in SARS-CoV-2 transmission. Lancet Infect Dis 2021;21(3):298−9. https://doi.org/10.1016/S1473-3099(20)30927-0.

[108] Goldstein E, Lipsitch M, Cevik M. On the effect of age on the transmission of SARS-CoV-2 in households, schools, and the community. J Infect Dis 2021;223(3):362−9. https://doi.org/10.1093/infdis/jiaa691.

[109] Calvani M, Cantiello G, Cavani M, Lacorte E, Mariani B, Panetta V, Parisi P, Parisi G, Roccabella F, Silvestri P, Vanacore N. Reasons for SARS-CoV-2 infection in children and their role in the transmission of infection according to age: a case-control study. Ital J Pediatr 2021;47(1):193. https://doi.org/10.1186/s13052-021-01141-1.

[110] Khalili M, Karamouzian M, Nasiri N, Javadi S, Mirzazadeh A, Sharifi H. Epidemiological characteristics of COVID-19: a systematic review and meta-analysis. Epidemiol Infect 2020;148:e130. https://doi.org/10.1017/S0950268820001430.

[111] Park M, Cook AR, Lim JT, Sun Y, Dickens BL. A systematic review of COVID-19 epidemiology based on current evidence. J Clin Med 2020;9(4). https://doi.org/10.3390/jcm9040967.

[112] Lessler J, Reich NG, Brookmeyer R, Perl TM, Nelson KE, Cummings DAT. Incubation periods of acute respiratory viral infections: a systematic review. Lancet Infect Dis 2009;9(5):291−300. https://doi.org/10.1016/S1473-3099(09)70069-6.

[113] McAloon C, Collins Á, Hunt K, Barber A, Byrne AW, Butler F, Casey M, Griffin J, Lane E, McEvoy D, Wall P, Green M, O'Grady L, More SJ. Incubation period of COVID-19: a rapid systematic review and meta-analysis of observational research. BMJ Open 2020;10(8):e039652. https://doi.org/10.1136/bmjopen-2020-039652.

[114] Wassie GT, Azene AG, Bantie GM, Dessie G, Aragaw AM. Incubation period of severe acute respiratory syndrome novel coronavirus 2 that causes coronavirus disease 2019: a systematic review and meta-analysis. Curr Ther Res Clin Exp 2020;93:100607. https://doi.org/10.1016/j.curtheres.2020.100607.

[115] Cheng C, Zhang D, Dang D, Geng J, Zhu P, Yuan M, Liang R, Yang H, Jin Y, Xie J, Chen S, Duan G. The incubation period of COVID-19: a global meta-analysis of 53 studies and a Chinese observation study of 11 545 patients. Infect Dis Poverty 2021a;10(1):119. https://doi.org/10.1186/s40249-021-00901-9.

[116] Kong T-K. More studies showing longer COVID-19 incubation period in older adults and questioning the appropriate times for quarantine and contact tracing. Aging Med (Milton (N.S.W)) 2020;3(4):278−9. https://doi.org/10.1002/agm2.12137.

[117] Guo C-X, He L, Yin J-Y, Meng X-G, Tan W, Yang G-P, Bo T, Liu J-P, Lin X-J, Chen X. Epidemiological and clinical features of pediatric COVID-19. BMC Med 2020;18(1):250. https://doi.org/10.1186/s12916-020-01719-2.

[118] She J, Liu L, Liu W. COVID-19 epidemic: disease characteristics in children. J Med Virol 2020;92(7):747−54. https://doi.org/10.1002/jmv.25807.

[119] Zhao C, Xu Y, Zhang X, Zhong Y, Long L, Zhan W, Xu T, Zhan C, Chen Y, Zhu J, Xiao W, He M. Public health initiatives from hospitalized patients with COVID-19, China. J Infect Public Health 2020;13(9):1229−36. https://doi.org/10.1016/j.jiph.2020.06.013.

[120] Zimmermann P, Curtis N. COVID-19 in children, pregnancy and neonates: a review of epidemiologic and clinical features. Pediatr Infect Dis J 2020;39(6):469–77. https://doi.org/10.1097/INF.0000000000002700.

[121] Wei WE, Li Z, Chiew CJ, Yong SE, Toh MP, Lee VJ. Presymptomatic transmission of SARS-CoV-2—Singapore, January 23–March 16, 2020. MMWR Morb Mort Wkly Rep 2020;69(14):411–5. https://doi.org/10.15585/mmwr.mm6914e1.

[122] Zou L, Ruan F, Huang M, Liang L, Huang H, Hong Z, Yu J, Kang M, Song Y, Xia J, Guo Q, Song T, He J, Yen H-L, Peiris M, Wu J. SARS-CoV-2 viral load in upper respiratory specimens of infected patients. N Engl J Med 2020;382(12):1177–9. https://doi.org/10.1056/NEJMc2001737.

[123] Byrne AW, McEvoy D, Collins AB, Hunt K, Casey M, Barber A, Butler F, Griffin J, Lane EA, McAloon C, O'Brien K, Wall P, Walsh KA, More SJ. Inferred duration of infectious period of SARS-CoV-2: rapid scoping review and analysis of available evidence for asymptomatic and symptomatic COVID-19 cases. BMJ Open 2020;10(8):e039856. https://doi.org/10.1136/bmjopen-2020-039856.

[124] Khera N, Santesmasses D, Kerepesi C, Gladyshev VN. COVID-19 mortality rate in children is U-shaped. Aging 2021;13(16):19954–62. https://doi.org/10.18632/aging.203442.

[125] Oliveira EA, Colosimo EA, Simões E Silva AC, Mak RH, Martelli DB, Silva LR, Martelli-Júnior H, Oliveira MCL. Clinical characteristics and risk factors for death among hospitalised children and adolescents with COVID-19 in Brazil: an analysis of a nationwide database. Lancet Child Adolesc. Health 2021;5(8):559–68. https://doi.org/10.1016/S2352-4642(21)00134-6.

[126] Dhir SK, Kumar J, Meena J, Kumar P. Clinical features and outcome of SARS-CoV-2 infection in neonates: a systematic review. J Trop Pediatr 2021;67(3). https://doi.org/10.1093/tropej/fmaa059.

[127] Liguoro I, Pilotto C, Bonanni M, Ferrari ME, Pusiol A, Nocerino A, Vidal E, Cogo P. SARS-COV-2 infection in children and newborns: a systematic review. Eur J Pediatr 2020;179(7):1029–46. https://doi.org/10.1007/s00431-020-03684-7.

[128] Hendler JV, Miranda do Lago P, Müller GC, Santana JC, Piva JP, Daudt LE. Risk factors for severe COVID-19 infection in Brazilian children. Braz J Infect Dis 2021:101650. https://doi.org/10.1016/j.bjid.2021.101650.

[129] Marsico C, Capretti MG, Aceti A, Vocale C, Carfagnini F, Serra C, et al. Severe neonatal COVID-19: challenges in management and therapeutic approach. J Med Virol 2021;94(4):1701–6. https://doi.org/10.1002/jmv.27472.

[130] Abrams EM, Sinha I, Fernandes RM, Hawcutt DB. Pediatric asthma and COVID-19: the known, the unknown, and the controversial. Pediatr Pulmonol 2020;55(12):3573–8. https://doi.org/10.1002/ppul.25117.

[131] Castro-Rodriguez JA, Forno E. Asthma and COVID-19 in children: a systematic review and call for data. Pediatr Pulmonol 2020;55(9):2412–8. https://doi.org/10.1002/ppul.24909.

[132] Green I, Merzon E, Vinker S, Golan-Cohen A, Magen E. COVID-19 susceptibility in bronchial asthma. J Allergy Clin Immunol Pract 2021;9(2):684–692.e1. https://doi.org/10.1016/j.jaip.2020.11.020.

[133] Lieberman-Cribbin W, Rapp J, Alpert N, Tuminello S, Taioli E. The impact of asthma on mortality in patients with COVID-19. Chest 2020;158(6):2290–1. https://doi.org/10.1016/j.chest.2020.05.575.

[134] Chhiba KD, Patel GB, Vu THT, Chen MM, Guo A, Kudlaty E, Mai Q, Yeh C, Muhammad LN, Harris KE, Bochner BS, Grammer LC, Greenberger PA, Kalhan R,

Kuang FL, Saltoun CA, Schleimer RP, Stevens WW, Peters AT. Prevalence and characterization of asthma in hospitalized and nonhospitalized patients with COVID-19. J Allergy Clin Immunol 2020;146(2):307–314.e4. https://doi.org/10.1016/j.jaci.2020.06.010.

[135] Choi H-G, Wee JH, Kim SY, Kim J-H, Il Kim H, Park J-Y, Park S, Il Hwang Y, Jang SH, Jung K-S. Association between asthma and clinical mortality/morbidity in COVID-19 patients using clinical epidemiologic data from Korean Disease Control and Prevention. Allergy 2021;76(3):921–4. https://doi.org/10.1111/all.14675.

[136] Lovinsky-Desir S, Deshpande DR, De A, Murray L, Stingone JA, Chan A, Patel N, Rai N, DiMango E, Milner J, Kattan M. Asthma among hospitalized patients with COVID-19 and related outcomes. J Allergy Clin Immunol 2020;146(5):1027–1034.e4. https://doi.org/10.1016/j.jaci.2020.07.026.

[137] Zhu Z, Hasegawa K, Ma B, Fujiogi M, Camargo CA, Liang L. Association of asthma and its genetic predisposition with the risk of severe COVID-19. J Allergy Clin Immunol 2020b;146(2):327–329.e4. https://doi.org/10.1016/j.jaci.2020.06.001.

[138] Halpin DMG, Singh D, Hadfield RM. Inhaled corticosteroids and COVID-19: a systematic review and clinical perspective. Eur Respir J 2020;55(5). https://doi.org/10.1183/13993003.01009-2020.

[139] Emes D, Hüls A, Baumer N, Dierssen M, Puri S, Russell L, Sherman SL, Strydom A, Bargagna S, Brandão AC, Costa ACS, Feany PT, Chicoine BA, Ghosh S, Rebillat A-S, Sgandurra G, Valentini D, Rohrer TR, Levin J, On Behalf Of The Trisomy Research Society Covid-Initiative Study Group. COVID-19 in children with Down syndrome: data from the Trisomy 21 Research Society Survey. J Clin Med 2021;10(21). https://doi.org/10.3390/jcm10215125.

[140] Soleimani A, Soleimani Z. Presentation and outcome of congenital heart disease during Covid-19 pandemic: a review. Curr Probl Cardiol 2021:100905. https://doi.org/10.1016/j.cpcardiol.2021.100905.

[141] Huang Y, Lu Y, Huang Y-M, Wang M, Ling W, Sui Y, Zhao H-L. Obesity in patients with COVID-19: a systematic review and meta-analysis. Metab Clin Exp 2020b;113:154378. https://doi.org/10.1016/j.metabol.2020.154378.

[142] Cheng WA, Turner L, Marentes Ruiz CJ, Tanaka ML, Congrave-Wilson Z, Lee Y, et al. Clinical manifestations of COVID-19 differ by age and obesity status. Influenza Other Respir Viruses 2021;16(2):255–64. https://doi.org/10.1111/irv.12918.

[143] Zhang F, Xiong Y, Wei Y, Hu Y, Wang F, Li G, Liu K, Du R, Wang C-Y, Zhu W. Obesity predisposes to the risk of higher mortality in young COVID-19 patients. J Med Virol 2020;92(11):2536–42. https://doi.org/10.1002/jmv.26039.

Chapter 2

COVID-19 and pediatrics—phylogeny, pathology, and pathogenesis of SARS-CoV-2

Eduard Matkovic[1] and Jessica Gulliver[1,2]
[1]*Department of Pathology and Laboratory Medicine, University of Wisconsin School of Medicine and Public Health, Madison, WI, United States;* [2]*Department of Pathology and Laboratory Medicine, Ann & Robert H. Lurie Children's Hospital of Chicago, Chicago, IL, United States*

Introduction

As the coronavirus disease 2019 (COVID-19) pandemic, caused by the spread of SARS-CoV-2, resulted in a global health emergency, the health and well-being of children and adolescents was significantly impacted through both COVID-19 disease and measures designed to contain the disease. Children living in vulnerable situations or with underlying medical conditions appear most likely to be affected. Since the beginning of the outbreak, the disease disproportionately affects the elderly, as fewer cases of COVID-19 are reported worldwide in children than in adults. Although a similar set of symptoms is shared between the adult and pediatric population, and people under the age of 20 years are susceptible to severe infection, the vast majority of pediatric cases are asymptomatic or mild [1,2]. Children and adolescents have a very low risk of developing severe disease, and treatment generally consists of supportive care. However, critical cases in children and adolescents result in hospital admission in the COVID-19 infection wards, and pediatric deaths are also reported [3,4].

As Omicron, classified by the World Health Organization (WHO) as a highly transmissible variant of concern on November 26, 2021, swept across the globe, children made up a significant proportion of SARS-CoV-2 positive cases compared to other periods during the pandemic. Millions of people were hospitalized and children were no exception, with increased numbers of COVID-19 pediatric hospitalizations [5]. Omicron seemed less likely to cause severe illness, hospitalization, or death compared to Delta; however, the high

Clinical Management of Pediatric COVID-19. https://doi.org/10.1016/B978-0-323-95059-6.00001-2

transmissibility coupled with lack of built up immunity left children vulnerable. With the potential for emergence of novel variants of concern, understanding the pathology and unique pathophysiological mechanisms that cause COVID-19 to disproportionately affect adults more than children is paramount. Compared to children without underlying disease, children or adolescents with comorbidities are susceptible to worse clinical outcomes and in rare cases, severe disease may lead to mortality [6]. The pre-existing conditions in the pediatric population such as childhood obesity, cardiac disease, and respiratory disease are prominent risk factors. While differences in outcomes between children and adults are not entirely clear, contributing factors that may play a role in why children fare better will be discussed in this review, with emphasis placed on the immune response, expression of the viral receptor, and presence of comorbidities. Furthermore, we will discuss the phylogeny and viral factors such as the pathology and pathogenesis associated with clinical disease in children and adolescents, highlighting mechanisms linked to children in high-risk groups.

Phylogeny and virology

SARS-CoV-2 is an enveloped, positive sense single-stranded RNA virus, a part of the genus betacoronavirus in the family *coronaviridae*. Coronaviruses possess one of the largest genomes among all RNA viruses and compared to other viruses, like influenza, the mutation rate is relatively slow as a result of the proofreading function and removal of mismatch nucleotides during replication. This taxonomic classification derives its name from the "crown-like" appearance visible by electron microscopy that is created by the surface spike (S) glycoprotein on the viral envelope. SARS-CoV-2 is closely related to two bat-derived coronaviruses, bat-SL-CoVZC45 and bat-SL-CoVZXC21, with 98% sequencing identity, and its genome shares approximately 79% and 50% sequence identity with SARS-CoV-1 and Middle East respiratory syndrome–related coronavirus (MERS-CoV), respectively [7]. Given the documented epidemiological history of the virus as well as evidence from genomic structure and ongoing evolution of SARS-CoV-2, there is a substantial body of scientific evidence pointing to a zoonotic event as the most parsimonious explanation of the origin of SARS-CoV-2 [8].

Pathogenesis of SARS-CoV-2

Coronavirus disease 2019 can be caused by the virus directly or through an excessive but ineffective host immune response. The mechanism by which SARS-CoV-2 directly causes disease is a two-step process that includes viral entry followed by replication within the respiratory tract. The first step for infection involves cellular entry, SARS-CoV-2 invades the respiratory epithelial cells by binding to angiotensin-converting enzyme 2 (ACE2) receptor with high affinity through the S protein [9]. Viral fusion with the cellular membrane is facilitated by conformational changes of the spike protein, which also undergoes proteolytic cleavage executed by serine and furin proteases [10]. Angiotensin-converting enzyme 2 (ACE-2) is an enzyme involved in maintaining homeostasis through the renin-angiotensin-system (RAS) by converting angiotensin 2 to angiotensin-(1−7) form [11]. Angiotensin-2 has properties involved in regulating vasoconstriction, inflammation, and fibrosis, while the forms of angiotensin-(1−7) counteract these effects. The cellular expression of ACE-2 receptor is downregulated by coronaviruses resulting in increased angiotensin-2 levels, and elevated angiotensin-2 increases vascular permeability and inflammation [12,13]. ACE-2 levels decrease with age and with hypertension or diabetes, which may explain why higher levels of ACE-2 activity in children may provide a protective role. However, age is not associated with ACE, ACE-2, and ratio of ACE-2:ACE activity in the lungs in patients with acute respiratory distress syndrome (ARDS) [14].

Replication of virus has been confirmed in areas of the lung and trachea that show histopathologic changes of acute lung injury, suggesting the virus directly induces injury and inflammation [15,16]. A transient viremia ensues after multiple iterations of viral replication causing clinical manifestation of classic "flu-like" symptoms to occur, similar to many other respiratory viruses. In addition to epithelial cells of airways, other cellular targets of SARS-CoV-2 include endothelial cells. Infection in the endothelial cells supports the hypothesis that SARS-CoV-2 may cause vascular dysfunction resulting in hypercoagulation and thrombosis. Activation of endothelial cells and immune cells leads to inflammation, and inflammation is closely integrated with the coagulation system. As cells are damaged, there is a release of cellular contents and inflammatory mediators, which is strongly prothrombotic. Eventually, the immune system in normal hosts should eliminate the virus through a humoral- and cell-mediated response. A robust innate immune system followed by an optimally functioning adaptive immune response in the overwhelming majority of healthy individuals is capable of suppressing viral replication and cellular invasion and resolving the infection. However, host factors do contribute to the susceptibility of severe infections particularly in the elderly (>65 years of age) and in individuals with multiple underlying health conditions, such as hypertension, obesity, chronic obstructive pulmonary disease, cardiovascular disease, and immunological disorders that result in an unbalanced host response. These predisposing factors seem to ease the progression of infection to severe disease sequelae that include ARDS, septic shock with secondary bacterial infection, coagulopathy, and hypercoagulability with formation of microvascular thrombi, refractory metabolic acidosis, and multiple organ failure. Development of the systemic inflammatory response syndrome is a serious condition stimulated by a maladaptive host immune response that generates excessive release of proinflammatory cytokines, referred to as the cytokine storm.

Obesity plays a role in pathogenesis of severe disease in pediatric COVID-19. Obesity and type 2 diabetes are associated with leptin deficiency or resistance. Leptin, an adipocyte-derived hormone, regulates food intake and energy hemostasis. By binding of cytokine type receptors, leptin induces proliferation of antigen presenting cells, monocytes, and T-helper cells, followed by subsequent secretion of proinflammatory cytokines, such as TNF-alpha, IL-2, and IL-6. Dysregulation of these mediators leads to a cytokine storm in obese patients infected with SARS-CoV-2 [17]. The pathogenesis in COVID-19 is quite complex and still under rigorous scientific investigation, as well as the immunobiology that is associated with a protracted or severe disease course. In children particularly, an abnormal immune response may lead to persistent tissue injury that is responsible for more harm to the host then the virus itself. Immunocompromised adults with malignancy and solid organ or hematopoietic stem cell transplant may be at increased risk of severe COVID-19 disease; however, according to some studies, immunocompromised pediatric patients are at no increased risk of severe disease [18–21].

Tissue injury is due in part to inflammation, vascular injury, and coagulopathy. In general, the pathologic features of COVID-19 are more or less identical to those seen in MERS-CoV and SARS-CoV-1 infections, although majority of histopathologic findings attributable to COVID-19 infections stem from adult autopsy reports. Nevertheless, children with comorbidities or infants are at greater risk for severe disease. Clinical issues that are associated with pediatric pathology include respiratory and neurologic manifestations, dermatologic findings, cardiovascular complications, and multisystemic inflammatory syndrome in children (MIS-C).

Pediatric COVID-19 typically has an uncomplicated clinical course, primarily characterized by mild respiratory symptoms or asymptomatic infection. As cases are not clinically detected, they may go undiagnosed. Why children with SARS-CoV-2 infection fare better than adults is obscure. One potential reason includes having a variety of memory T-cells from exposure to different viruses at a young age, which primes the immune system to cross-react with SARS-CoV-2. As children develop so does the immune system, and therefore, it is possible that an immature immune defense is less likely to cause an excessive host response that results in an inflammatory state with a cytokine surge. Lastly, the immune mediated effect in young people after being vaccinated is better than that of the elderly who have a waning immunity known as immunosenescence.

Multisystemic inflammatory syndrome in children

Children develop a postinflammatory response known at MIS-C following infection with SARS-CoV-2. A direct causal relationship with SARS-CoV-2 and MIS-C has yet to be established; however, there is abundant epidemiologic data linking the virus to the syndrome. The syndrome manifests approximately

2–4 weeks following infection. It is thought to be a postinfectious mechanism rather than due to acute COVID-19 infection as there is a temporal association with a lag phase of MIS-C cases that follows a surge of SARS-CoV-2 positive cases. The prevailing theory purports three overlapping phases of disease with the first being an acute asymptomatic infection. The second phase is a pulmonary phase that is more severe in adults but mild in children, followed by the third phase—a delayed immune response with overt signs of systemic inflammatory response and multiple organ dysfunction and failure. The majority of children with MIS-C are healthy without any underlying health conditions as was described with severe COVID-19 infection. While the exact incidence is unknown as many cases are mild or go unreported, there are more severe cases with a broad range of clinical features that occur in children between 6 and 12 years of age. The first reported cases came out of the United Kingdom in late April 2020 in previously healthy children presenting with cardiovascular shock, fever, and hyperinflammation [22]. Although there is no single pathognomonic finding to diagnosis MIS-C, according to the Centers of Disease and Prevention the case definition for MIS-C associated with SARS-CoV-2 includes: a person <21 years of age requiring hospitalization for a clinically severe illness with fever, laboratory evidence of inflammation and multiple organ involvement, and confirmed SARS-CoV-2 infection or exposure to a confirmed COVID-19 case (see Table 2.1) [23].

TABLE 2.1 The case definition for multisystem inflammatory syndrome in children requires a clinicopathological correlation, as described by the Centers for Disease Control and Prevention (CDC) Health Advisory. Consideration of multisystemic inflammatory syndrome in children (MIS-C) should be given to any pediatric fatality with evidence of SARS-CoV-2 infection.

Case definition for multisystem inflammatory syndrome in children Associated with coronavirus disease 2019
Individual aged <21 years presenting with fever (>38.0°C for ≥24 h or report of subjective symptoms of fever lasting ≥24 h), evidence of clinically severe illness require hospitalization with multisystem (>2) organ involvement (cardiac, renal, respiratory, hematologic, gastrointestinal, dermatologic or neurological), and no alternative plausible diagnosis; AND
Laboratory evidence of inflammation (including but not limited to one or more of the following; elevated CRP, ESR, fibrinogen, procalcitonin, d-dimer, ferritin, LDH or interleukin 6 (IL-6), neutrophilia, lymphopenia, and hypoalbuminemia; AND
Positive for current or recent SARS-CoV-2 infection by RT-PCR, serology, or antigen test; or exposure to a suspected or confirmed COVID-19 case within the 4 weeks prior to the onset of symptoms

The clinical presentation may resemble Kawasaki's disease, toxic shock syndrome (TSS), or secondary hemophagocytic lymphohistiocytosis (HLH). The patients present with symptoms of fever, mucocutaneous rash, abdominal pain, lymphadenopathy, myocarditis, ileitis, or edema in the upper and lower limbs. The hyperinflammatory state can lead to exudates with ascites, pleural, and pericardial effusions, and additionally, a hyperferritinemia syndrome can occur in tandem with lymphopenia, commonly seen in HLH-related disorders [24]. Superantigen-like motifs recognized in the protein structure of SARS-CoV-2 may explain the clinical presentation resembling TSS, with diffuse erythrodermic rash and severe hypotension as originally seen with *Staphylococcus aureus* and *Streptococcus pyogenes* [25]. Elevated inflammatory markers such as erythrocyte sedimentation rate (ESR), procalcitonin, and C-reactive protein (CRP) may occur with signs of tissue injury and organ failure as indicated by elevated levels of lactate dehydrogenase (LDH), liver transaminases, blood urea nitrogen (BUN) in kidney injury, and troponin in cardiac injury. Laboratory markers that are positively correlated with severity and mortality include a decreased CD4(+) and CD8(+) T cell response, high LDH, and prominent neutrophilia in the peripheral blood (see Fig. 2.1).

Whereas direct tissue injury from viral-induced cytopathic effect is less likely to contribute to MIS-C, the cellular response is driven by the cytokine production, involving monocytes and stimulated macrophages, neutrophils, and B-cells. As a part of the humoral response, plasma cells begin to synthesize antibodies specific for the viral epitopes such as the spike protein or nucleocapsid. Macrophage activation triggers cytokine release and stimulation of T-helper cells and various other immune cells, which further amplifies the

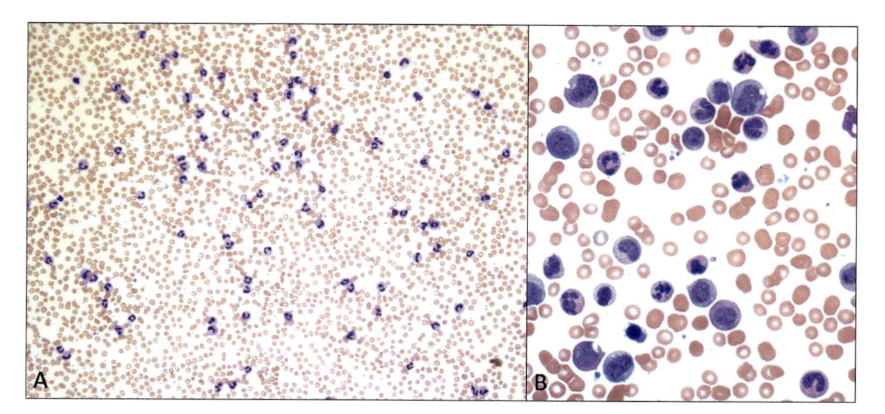

FIGURE 2.1 Peripheral blood with markedly left shifted neutrophilia with occasional atypical lymphocytes in patients with severe acute respiratory syndrome–associated coronavirus 2 (SARS-CoV-2) infection. The white cell count was 50,000 cells/mcL with differential count of 88% neutrophils and 1% lymphocytes. The morphologic findings may mimic chronic myeloid leukemia except for the absence of basophilia (A). Prominent left shift noted on higher power with no features of dysplasia and occasional reactive atypical lymphocytes cells seen (B).

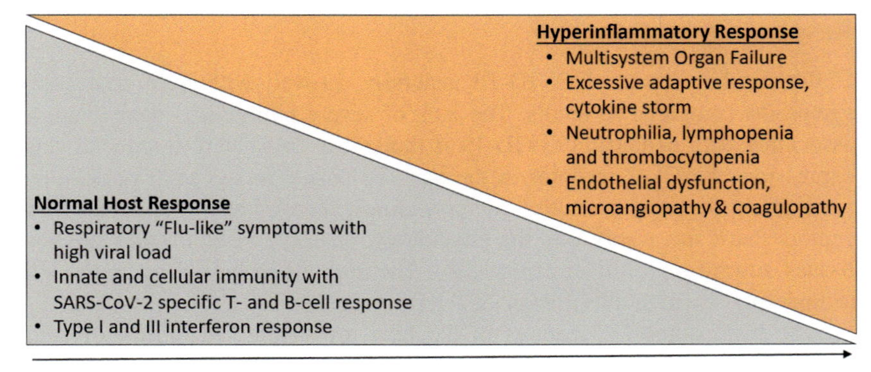

FIGURE 2.2 Pathogenesis of pediatric coronavirus disease 2019 (COVID-19) with milder disease symptoms than in adults compared with multisystem inflammatory syndrome in children with vascular complications and shock.

cytokine milieu. Ultimately, the hyperinflammatory response becomes dysregulated and unconstrained (see Fig. 2.2). Neutrophilia, lymphopenia, and thrombocytopenia are commonly found in MIS-C and may be regarded as markers of disease severity (see Fig. 2.2) [26]. Neutrophil activation during infection coupled with elevated cytokines, and endothelial dysfunction may contribute to the increased inflammatory state and severe clinical course [26]. Lymphopenia in pediatric COVID-19 and MIS-C shows different patterns of immune activation likely attributed to different subsets of T-lymphocyte populations [25]. In the case of MIS-C, the T-cell immune response may be a useful marker for development of shock in severe MIS-C cases [27,28]. Atypical lymphocytes in the peripheral blood appear after disease onset in patients who are more likely to have pneumonia [29].

Kawasaki disease is an acute medium vessel vasculitis that has features of rash, cervical lymphadenopathy, rash of ocular or oral mucosa, and acquired heart disease in the form of coronary artery aneurysm. Unlike Kawasaki's disease, MIS-C predominantly affects slightly older children and is associated with more frequent cardiovascular involvement [30]. Dermatologic symptoms in MIS-C are highest among children ages 0–5 years, whereas the prevalence of myocarditis is highest among adolescents. Cardiovascular symptoms are relatively common in this MIS-C cohort and may be due to inflammatory myocarditis or coronary artery dilation, which tends to occur in a younger patient population similar to Kawasaki's disease. The pathogenesis of MIS-C is not fully understood but is considered to be due to a cytokine storm derived from the coronaviruses ability to cause expression of cytokines and chemokines and block certain interferon responses in the host [31]. The cytokine storm leads to tissue dysfunction with elevated enzymes and eventual organ failure.

Respiratory system

Children who contract COVID-19 generally do well with mild respiratory symptoms compared to adults. The lack of severe pulmonary manifestations associated with pediatric COVID-19 is thought to be related to reduced gene expression of ACE-2 receptor in the targeted tissue. In the adult population, particularly those with multiple comorbidities, there is an excess ACE-2 receptors that conceivably tilts the physiology, once the virus invades the host tissues, toward a proinflammatory state. The immune response is also thought to underpin the pathophysiology of ARDS.

The histopathologic features in the lung are comprised mostly of adult autopsy cases, although it is postulated that pediatric patients with severe COVID-19 requiring mechanical ventilation could have similar lung pathology (see Fig. 2.3). The predominant pattern of lung injury is of diffuse alveolar damage (DAD), the histopathologic correlate of ARDS, and its evolving stages [16]. The acute (or exudative) phase develops within the first week of acute lung injury, of any kind, and is characterized by neutrophilic inflammation that increases the permeability of airway spaces that causes intra-alveolar hemorrhage and leakage of protein rich edema in the alveolar spaces. There is a decrease in alveolar compliance with epithelial denudation and hyaline membrane formation. The early phase is followed by a proliferative (organizing) phase that corresponds to a subacute clinical course and histologically

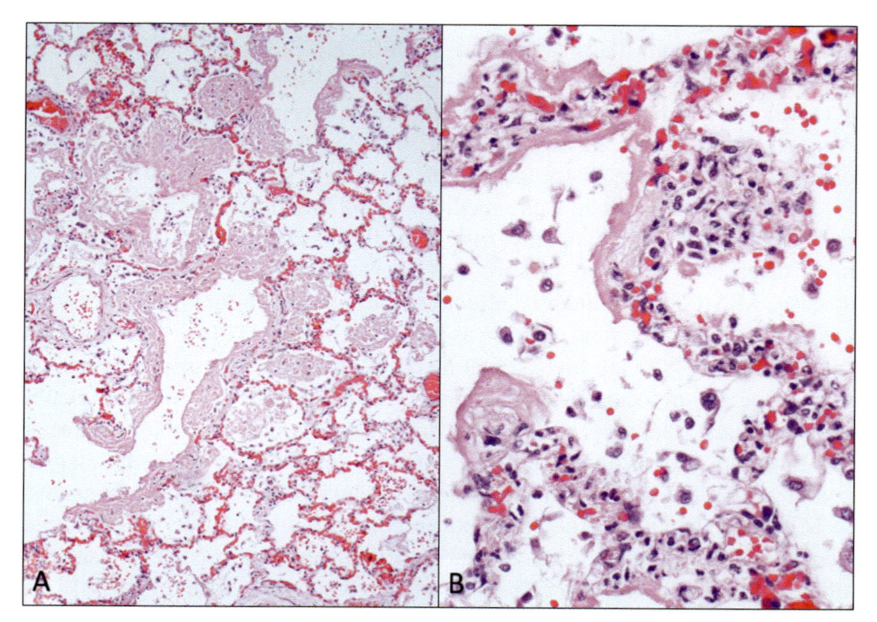

FIGURE 2.3 Lung specimen with acute phase of diffuse alveolar damage with fibrinous exudates (A). Higher power view demonstrates acute interstitial inflammatory infiltrate with hyaline membrane formation (B).

shows type II pneumocyte hyperplasia, proliferating fibroblasts, fibrin deposits, and a lymphocytic infiltrate. While no viral inclusions are seen, nonspecific changes suggestive of viral infection are noted in the form of multinucleated enlarged pneumocytes with large nuclei in the alveolar spaces. Occasionally, a pattern of acute fibrinous and organizing pneumonia may be seen, characterized by marked fibrin deposition. Perivascular inflammation can be seen, and pulmonary vasculature may show microvascular damage with microthrombi that is linked to a prothrombotic state. The tissue microenvironment contains an excess of inflammatory cytokines that are released by the immune cellular infiltrate. In the course of time, a dense collagenous fibrosis remodels the lung's histologic architecture, known as the fibrotic phase. A secondary bacterial pneumonia can occur shortly following infection of SARS-CoV-2 and complicate the clinical picture. Histology shows a characteristic acute neutrophilic pneumonia in this scenario (Fig. 2.4). Viral antigens, observed in tissues using immunohistochemistry, are identified in upper and lower respiratory tract tissues in the lung parenchyma [16]. In situ hybridization using RNA probes is a reliable diagnostic methodology used to demonstrate viral tropism and replication within tissue sections in tracheal epithelium and epithelial cells in airways deep in the lung parenchyma [15].

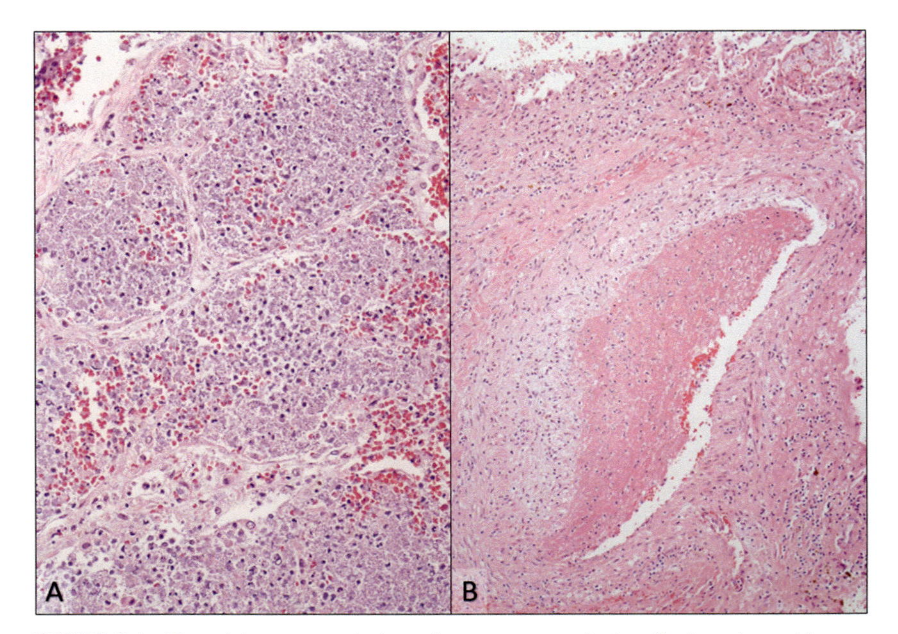

FIGURE 2.4 Necrotizing pneumonia in an immunocompromised pediatric patient with coronavirus disease 2019 (COVID-19) complicated by a superinfection with *Pseudomonas spp* (A). The same patient had aspergillosis involving the central nervous system. A thrombus in a pulmonary vessel was detected; however, no fungal organisms were identified in the vessel wall with special stains (Grocott's methenamine silver and periodic acid−Schiff stains) suggesting the pathology was due to COVID-19 complicated by pseudomonas infection.

These histopathologic findings in tandem with the rate at which DAD occurs were similar to other observed pathology described with H1N1 influenza and SARS [32,33]. Microthrombi were reported at higher rates with COVID-19 compared with H1N1, suggesting that SARS-CoV-2 may have a predilection for endothelial cells resulting in an endotheliopathy that could increase the incidence of hypoxic complications due to thrombi [33]. This may also be a byproduct of the immune naïve population and the emergence of a novel infectious disease causing an excessive immunologic response to the viral infection. Direct infection of vascular endothelium has been confirmed by localizing SARS-CoV-2 RNA in endothelial cells and tunica media of vessels in multiple tissues, including lungs, brain, heart, liver, and kidney [15]. In addition to microthrombi, pulmonary thrombosis in larger vessels has been reported in COVID-19 autopsy patients in adults but has only rarely been diagnosed in the pediatric setting (Fig. 2.4) [33,34].

A rising infection rate in pediatric patients coincided with the rapid community spread of the SARS-CoV-2 omicron (B1.1.529) variant. Among those admitted to a hospital, fever, cough, and shortness of breath were the more common symptoms reported and rarely, patients required ventilation [5]. Overall, COVID-19 viral pneumonia in children is mainly mild. The majority of these cases do not undergo tissue biopsy as early chest CT shows characteristic changes that can be used for diagnosis and monitoring. In the case of co-infection, additional microbiologic testing would be indicated to assist clinical management. Histopathologic features in patients with asymptomatic SARS-CoV-2 have been described in adults and may recapitulate the pathology in pediatric cases [33]. Unsurprisingly, asymptomatic cases had less severe histopathologic findings compared to symptomatic cases with features of proteinaceous exudates, pneumocyte hyperplasia, and patchy chronic infiltrates with scattered multinucleated cells. Asymptomatic cases lack prominent hyaline membranes or marked fibrinous exudates with acute inflammation; viral inclusions were not present. Bronchial alveolar lavages (BALs) preformed on children at our institution showed nonspecific findings on cytology specimens that included increased hemosiderin laden macrophages (see Fig. 2.5). This finding is nonspecific and commonly associated with other lung pathologies such as acute lung injury in the setting of e-cigarette use or vaping, which should be included on the differential diagnosis as it has similar diagnostic features to COVID-19 and also, e-cigarette smokers have high rates of SARS-CoV-2 infection [35,36].

Cardiovascular system

While COVID-19 is a disease of the respiratory system, the cardiovascular system is considerably affected during acute infection as well as in the convalescent phase. Several factors can cause lasting damage to the heart

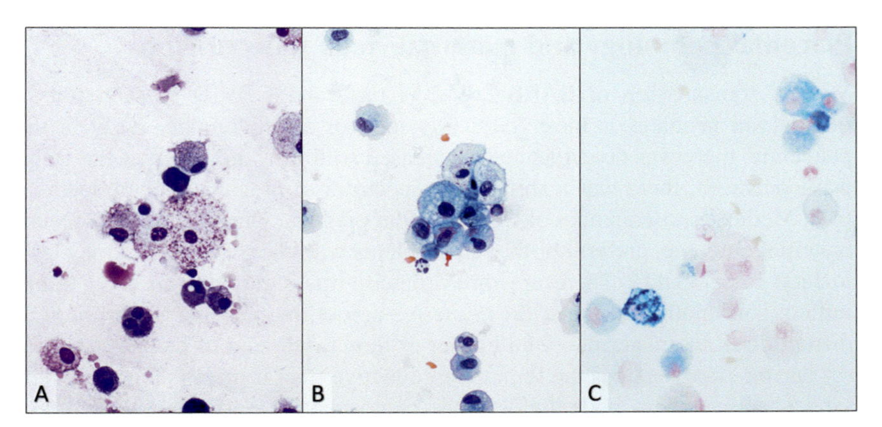

FIGURE 2.5 Cytology specimens from a bronchial alveolar lavage (BAL) of a pediatric patient with coronavirus disease 2019 (COVID-19) demonstrating foamy clusters of macrophages containing lipid contents (confirmed on positive staining with Oil Red O special stain) (A and B). Iron stain shows the macrophages are laden with hemosiderin (C). The cytology in pediatric COVID-19 is nonspecific and may be seen in other conditions of acute lung injury. A broad differential would be needed, and morphologic findings should be correlated with radiographic and other ancillary laboratory tests.

tissue such as hypoxia, myocarditis, and vasculitis. Catecholamine excess may also be a key mediator deleteriously altering the regulatory physiologic response, as well as complement-mediated vascular injury [37,38]. COVID-19 has expanded the list of viral infections known to cause myocarditis in children, as fulminant myocarditis cases have been reported [39]. The pathology characteristically shows a myeloid rich inflammatory infiltrate in the epicardium and myocardium without myocyte necrosis; associated myocyte hypertrophy is present. Myocarditis can progress to a dilated cardiomyopathy, which may greatly affect the quality of life even after the inflammation has subsided. The pathogenesis of COVID-19-induced myocarditis is most likely an immune-mediated host response bringing about a proinflammatory cytokine surge, microangiopathy with thromboinflammation, and endothelial dysfunction. However, there is evidence of direct cytotoxic tissue damage via ACE2 mediated cardiomyocyte entry [40]. Children with cardiac disease have an increased risk of critical illness that poorly affects the clinical course, especially neonates and infants with a congenital heart disease. Although a subset of pediatric patients may not have underlying health conditions, similar to the healthy phenotype seen frequently in MIS-C. While myopericarditis post-mRNA COVID-19 vaccination in adolescents has been described, the clinical presentation is associated with a much milder hospital course than COVID-19-associated myocarditis [41,42]. Moreover, the overall risk of cardiac or other noncardiac complications of SARS-CoV-2 infection is higher than in vaccine-related myocarditis [42].

Placental pathology and maternal–fetal transmission

Vertical transmission of SARS-CoV-2 is rare, and COVID-19 infection in infants and neonates is most frequently mild or asymptomatic. Early in the pandemic, there was no evidence of increased stillbirths; however, as the Delta wave occurred, there was a shift in the spectrum of pregnancy complications [43]. Medical consequences of COVID-19 in pregnancy increase risk of severe maternal outcome, preterm birth, and problems with the placenta that may lead to fetal loss [43,44]. There are three routes viruses can be vertically transmitted from mother to fetus: (a) in utero—across the placenta either through disruption in the maternal–fetal barrier or direct infection of placental tissue; (b) during labor—across the female reproductive tract if there is high levels of viral shedding from body fluids; and (c) postnatal breast feeding. In utero SARS-CoV-2 transmission has been confirmed with evidence of localized viral SARS-CoV-2 RNA in vascular endothelium in heart and liver of neonatal autopsy cases where the mothers had asymptomatic SARS-CoV-2 infection [45]. Syncytiotrophoblast necrosis, increased perivillious fibrin, and intervillositis are some of the well-described pathologic features of SARS-CoV-2 placental infections [43,46,47]. The placenta can show increased vascularity, decidual arteriopathy, and chronic inflammation with plasma cells as well as features of both fetal and maternal vascular malperfusion. There is a broad pattern of histopathologic features associated with SARS-CoV-2 infection suggesting multiple pathogenic factors are involved, ranging from systemic inflammation, hypoxia-induced changes, and coagulopathy. Importantly, some histopathologic features are more frequently observed in cases with adverse perinatal outcomes, although robust epidemiologic studies are needed to corroborate the pathologic findings and assess the overall obstetric health outcomes [43,47].

Dermatopathology

There is a wide range of cutaneous manifestations of COVID-19 that have been reported from a number of cases from around the world [44,49]. Although presenting skin symptoms during infection are considered infrequent, COVID-19-associated skin lesions are probably underrecognized. It is unclear if these skin findings are clinically associated with illness severity [50,51]. Retiform purpura, a potential sign of coagulopathy, has been associated with severe disease and high viral load, primarily in adults [52]. The lesion appears as a branching, nonblanching plaque or patch with skin necrosis and ulceration due to ischemia. Chilblain-like lesion, colloquially referred to as "COVID-toes," are described in young children or adolescents, primarily on the feet, and in children who are generally in good health [48,52]. Skin biopsies of the acral skin shows a superficial and deep chronic perivascular inflammatory infiltrate composed of small mature lymphocytes with occasional scattered clusters of plasma cells [48]. Dermal biopsies performed on children with MIS-C have similar morphology including features of thrombotic microangiopathy with microthrombi present in papillary dermal capillaries. Given that the cutaneous manifestation peculiarly occur in colder weather, a differential diagnosis would include cryoglobulin and cold agglutinin disease. Some experts believe the "COVID-toe" may be something unrelated to the virus, possibly related to an immunologic phenomenon [53]. There are several other cutaneous manifestations including erythema multiforme, urticarial, and perniosis, which are considered reactions with significant type I interferon signaling [52].

Neuropathology

Neurologic disease from COVID-19 in children is rare, although vascular and neurologic complications have emerged as a serious complication from the highly pathogenic SARS-CoV-2 infection. Neuropathology has been well-defined in adults, but studies and even clinical case reports on children are infrequent. The pathogenic mechanism underlying the neurologic manifestations is complex and multifactorial. Brain tissue shows chronic and acute neurologic abnormalities in adult COVID-19 autopsy cases, with acute hypoxic−ischemic changes, hemorrhage, and minimal inflammation frequently observed and low levels of viral SARS-CoV-2 RNA present [54]. There is sound evidence from human adult cases showing viral infection and replication in vascular endothelial cells within the brain [15]. Animal models suggest viral entry into the central nervous system (CNS) occurs through the olfactory bulb followed by progressive invasion of subcortical structures [55]. The transneuronal route has not been confirmed in human autopsy cases. Other potential routes to penetrate into the CNS include hematogenous with high levels of viremia in the peripheral blood, invasion through the neuronal vasculature, or infection of monocytes/macrophage that serve as potential reservoirs, analogous to the Trojan horse [56].

Pediatric cases with histopathologic findings are scarce. There is a pediatric case described in literature of a 7-year-old boy who presented with 3 days of fever with headache and dizziness and confirmed SARS-CoV-2 positive on nasopharyngeal swab [57]. The patient's mental status began to fluctuate, and eventually he became unresponsive with loss of brainstem reflexes and dilated pupils. Electroencephalogram (EEG) was abnormal and head CT demonstrated multiple acute parenchymal hemorrhages involving the cerebral hemispheres bilaterally. Laboratory inflammatory markers, including procalcitonin, CRP, and D-dimers, were significantly elevated. The pathologic findings associated with the COVID-related encephalopathy included angiocentric mixed mononuclear inflammatory infiltrates, lymphohistocytic inflammation in leptomeninges, and shrunken and hyperchromatic neurons with edematous neuropil. Ultrastructural analysis, PCR, and immunohistochemistry for SARS-CoV-2 were all negative, suggesting the cerebral vasculitis was caused by an inflammatory state provoked in part by immune cell recruitment and cytokine secretion. Coagulation disorders also likely contribute to the vascular and neurologic injury.

Summary

The worldwide spread of pediatric SARS-CoV-2 infection justifies a global effort to understand the pathology of the disease and use that knowledge to implement successful public health strategies to reduce spread and optimize the medical management. Children of all ages can become infected with COVID-19 and experience its complications. While there is a spectrum of illness from SARS-CoV-2 infection in children, most children are not susceptible to severe disease. Children and adolescents experience mainly mild symptoms or have a completely asymptomatic infection, providing a potential infectious source of spread. However, certain children with underlying medical conditions may be at increased risk of severe disease. More serious conditions observed with pediatric COVID-19 include MIS-C, respiratory distress, myo/pericarditis and cerebral vasculitis, but fortunately, all these conditions are rare. There are distinct histopathologic features of COVID-19 in adults, but the lack of data on pediatric patients requires further pathology studies that need correlation with epidemiological patterns and clinical outcomes.

References

[1] Davies NG, Klepac P, Liu Y, Prem K, Jit M, Eggo RM. Age-dependent effects in the transmission and control of COVID-19 epidemics. Nat Med August 2020;26(8):1205−11.

[2] Castagnoli R, Votto M, Licari A, Brambilla I, Bruno R, Perlini S, Rovida F, Baldanti F, Marseglia GL. Severe acute respiratory syndrome coronavirus 2 (SARS-CoV-2) infection in children and adolescents: a systematic review. JAMA Pediatr September 1, 2020;174(9):882−9.

[3] Delahoy MJ, Ujamaa D, Whitaker M, O'Halloran A, Anglin O, Burns E, Cummings C, Holstein R, Kambhampati AK, Milucky J, Patel K. Hospitalizations associated with COVID-19 among children and adolescents—COVID-NET, 14 states, March 1, 2020–August 14, 2021. MMWR (Morb Mortal Wkly Rep) September 10, 2021;70(36):1255.

[4] McCormick DW, Richardson LC, Young PR, Viens LJ, Gould CV, Kimball A, Pindyck T, Rosenblum HG, Siegel DA, Vu QM, Komatsu K. Deaths in children and adolescents associated with COVID-19 and MIS-C in the United States. Pediatrics November 1, 2021;148(5).

[5] Cloete J, Kruger A, Masha M, du Plessis NM, Mawela D, Tshukudu M, Manyane T, Komane L, Venter M, Jassat W, Goga A. Paediatric hospitalisations due to COVID-19 during the first SARS-CoV-2 omicron (B. 1.1. 529) variant wave in South Africa: a multicentre observational study. The Lancet Child & Adolescent Health May 1, 2022;6(5):294−302.

[6] Tsankov BK, Allaire JM, Irvine MA, Lopez AA, Sauvé LJ, Vallance BA, Jacobson K. Severe COVID-19 infection and pediatric comorbidities: a systematic review and meta-analysis. Int J Infect Dis February 1, 2021;103:246−56.

[7] Lu R, Zhao X, Li J, Niu P, Yang B, Wu H, Wang W, Song H, Huang B, Zhu N, Bi Y. Genomic characterisation and epidemiology of 2019 novel coronavirus: implications for virus origins and receptor binding. Lancet February 22, 2020;395(10224):565–74.

[8] Holmes EC, Goldstein SA, Rasmussen AL, Robertson DL, Crits-Christoph A, Wertheim JO, Anthony SJ, Barclay WS, Boni MF, Doherty PC, Farrar J. The origins of SARS-CoV-2: a critical review. Cell September 16, 2021;184(19):4848–56.

[9] Walls AC, Park YJ, Tortorici MA, Wall A, McGuire AT, Veesler D. Structure, function, and antigenicity of the SARS-CoV-2 spike glycoprotein. Cell April 16, 2020;181(2):281–92.

[10] Peng R, Wu LA, Wang Q, Qi J, Gao GF. Cell entry by SARS-CoV-2. Trends Biochem Sci October 1, 2021;46(10):848–60.

[11] Guo J, Huang Z, Lin L, Lv J. Coronavirus disease 2019 (COVID-19) and cardiovascular disease: a viewpoint on the potential influence of angiotensin-converting enzyme inhibitors/ angiotensin receptor blockers on onset and severity of severe acute respiratory syndrome coronavirus 2 infection. J Am Heart Assoc April 9, 2020;9(7):e016219.

[12] De Wit E, Van Doremalen N, Falzarano D, Munster VJ. SARS and MERS: recent insights into emerging coronaviruses. Nat Rev Microbiol August 2016;14(8):523–34.

[13] Imai Y, Kuba K, Rao S, Huan Y, Guo F, Guan B, Yang P, Sarao R, Wada T, Leong-Poi H, Crackower MA. Angiotensin-converting enzyme 2 protects from severe acute lung failure. Nature July 2005;436(7047):112–6.

[14] Schouten LR, van Kaam AH, Kohse F, Veltkamp F, Bos LD, de Beer FM, van Hooijdonk RT, Horn J, Straat M, Witteveen E, Glas GJ. Age-dependent differences in pulmonary host responses in ARDS: a prospective observational cohort study. Ann Intensive Care December 2019;9(1):1–9.

[15] Bhatnagar J, Gary J, Reagan-Steiner S, Estetter LB, Tong S, Tao Y, Denison AM, Lee E, DeLeon-Carnes M, Li Y, Uehara A. Evidence of severe acute respiratory syndrome coronavirus 2 replication and tropism in the lungs, airways, and vascular endothelium of patients with fatal coronavirus disease 2019: an autopsy case series. J Infect Dis March 1, 2021;223(5):752–64.

[16] Martines RB, Ritter JM, Matkovic E, Gary J, Bollweg BC, Bullock H, Goldsmith CS, Silva-Flannery L, Seixas JN, Reagan-Steiner S, Uyeki T. Pathology and pathogenesis of SARS-CoV-2 associated with fatal coronavirus disease, United States. Emerg Infect Dis September 2020;26(9):2005.

[17] Maurya R, Sebastian P, Namdeo M, Devender M, Gertler A. COVID-19 severity in obesity: leptin and inflammatory cytokine interplay in the link between high morbidity and mortality. Front Immunol June 18, 2021;12:2349.

[18] Caillard S, Anglicheau D, Matignon M, Durrbach A, Greze C, Frimat L, Thaunat O, Legris T, Moal V, Westeel PF, Kamar N. An initial report from the French SOT COVID registry suggests high mortality due to COVID-19 in recipients of kidney transplants. Kidney Int December 1, 2020;98(6):1549–58.

[19] Sharma A, Bhatt NS, St Martin A, Abid MB, Bloomquist J, Chemaly RF, Dandoy C, Gauthier J, Gowda L, Perales MA, Seropian S. Clinical characteristics and outcomes of COVID-19 in haematopoietic stem-cell transplantation recipients: an observational cohort study. The Lancet Haematology March 1, 2021;8(3):e185–93.

[20] Chappell H, Patel R, Driessens C, Tarr AW, Irving WL, Tighe PJ, Jackson HJ, Harvey-Cowlishaw T, Mills L, Shaunak M, Gbesemete D. Immunocompromised children and young people are at no increased risk of severe COVID-19. J Infect January 1, 2022;84(1):31–9.

[21] El Dannan H, Al Hassani M, Ramsi M. Clinical course of COVID-19 among immunocompromised children: a clinical case series. BMJ Case Reports CP October 1, 2020;13(10):e237804.

[22] Riphagen S, Gomez X, Gonzalez-Martinez C, Wilkinson N, Theocharis P. Hyperinflammatory shock in children during COVID-19 pandemic. Lancet May 23, 2020;395 (10237):1607–8.

[23] Centers for Disease Control and Prevention CDC Health Advisory—Emergency Preparedness and Response, distributed via CDC Health Alert Network – accessed May 20, 2022 at https://emergency.cdc.gov/han/2020/han00432.asp.

[24] Mamishi S, Movahedi Z, Mohammadi M, Ziaee V, Khodabandeh M, Abdolsalehi MR, Navaeian A, Heydari H, Mahmoudi S, Pourakbari B. Multisystem inflammatory syndrome associated with SARS-CoV-2 infection in 45 children: a first report from Iran. Epidemiol Infect 2020:148.

[25] Cheng MH, Zhang S, Porritt RA, Rivas MN, Paschold L, Willscher E, Binder M, Arditi M, Bahar I. Superantigenic character of an insert unique to SARS-CoV-2 spike supported by skewed TCR repertoire in patients with hyperinflammation. Proc Natl Acad Sci USA October 13, 2020;117(41):25254–62.

[26] Vogel TP, Top KA, Karatzios C, Hilmers DC, Tapia LI, Moceri P, Giovannini-Chami L, Wood N, Chandler RE, Klein NP, Schlaudecker EP. Multisystem inflammatory syndrome in children and adults (MIS-C/A): case definition & guidelines for data collection, analysis, and presentation of immunization safety data. Vaccine May 21, 2021;39(22):3037–49.

[27] Vella LA, Giles JR, Baxter AE, Oldridge DA, Diorio C, Kuri-Cervantes L, Alanio C, Pampena MB, Wu JE, Chen Z, Huang YJ. Deep immune profiling of MIS-C demonstrates marked but transient immune activation compared with adult and pediatric COVID-19. Science immunology March 2, 2021;6(57):eabf7570.

[28] Okarska-Napierała M, Mańdziuk J, Feleszko W, Stelmaszczyk-Emmel A, Panczyk M, Demkow U, Kuchar E. Recurrent assessment of lymphocyte subsets in 32 patients with multisystem inflammatory syndrome in children (MIS-C). Pediatr Allergy Immunol November 2021;32(8):1857–65.

[29] Sugihara J, Shibata S, Doi M, Shimmura T, Inoue S, Matsumoto O, Suzuki H, Makino A, Miyazaki Y. Atypical lymphocytes in the peripheral blood of COVID-19 patients: a prognostic factor for the clinical course of COVID-19. PLoS One November 12, 2021;16(11):e0259910.

[30] Feldstein LR, Rose EB, Horwitz SM, Collins JP, Newhams MM, Son MB, Newburger JW, Kleinman LC, Heidemann SM, Martin AA, Singh AR. Multisystem inflammatory syndrome in US children and adolescents. N Engl J Med July 23, 2020;383(4):334–46.

[31] Blanco-Melo D, et al. Imbalanced host response to SARS-CoV-2 drives development of COVID-19. Cell 2020;181:1036–1045.e9.

[32] Shieh WJ, Blau DM, Denison AM, DeLeon-Carnes M, Adem P, Bhatnagar J, Sumner J, Liu L, Patel M, Batten B, Greer P. 2009 pandemic influenza A (H1N1): pathology and pathogenesis of 100 fatal cases in the United States. Am J Pathol July 1, 2010;177(1):166–75.

[33] Hariri LP, North CM, Shih AR, Israel RA, Maley JH, Villalba JA, Vinarsky V, Rubin J, Okin DA, Sclafani A, Alladina JW. Lung histopathology in coronavirus disease 2019 as compared with severe acute respiratory sydrome and H1N1 influenza: a systematic review. Chest January 1, 2021;159(1):73–84.

[34] Chima M, Williams D, Thomas NJ, Krawiec C. COVID-19–associated pulmonary embolism in pediatric patients. Hosp Pediatr June 1, 2021;11(6):e90–4.

[35] Reagan-Steiner S, Gary J, Matkovic E, Ritter JM, Shieh WJ, Martines RB, Werner AK, Lynfield R, Holzbauer S, Bullock H, Denison AM. Pathological findings in suspected cases of e-cigarette, or vaping, product use-associated lung injury (EVALI): a case series. Lancet Respir Med December 1, 2020;8(12):1219−32.
[36] Kazachkov M, Pirzada M. Diagnosis of EVALI in the COVID-19 era. Lancet Respir Med December 1, 2020;8(12):1169−70.
[37] Magro C, Mulvey JJ, Berlin D, Nuovo G, Salvatore S, Harp J, Baxter-Stoltzfus A, Laurence J. Complement associated microvascular injury and thrombosis in the pathogenesis of severe COVID-19 infection: a report of five cases. Transl Res June 1, 2020;220:1−3.
[38] Gubbi S, Nazari MA, Taieb D, Klubo-Gwiezdzinska J, Pacak K. Catecholamine physiology and its implications in patients with COVID-19. Lancet Diabetes Endocrinol December 1, 2020;8(12):978−86.
[39] Lara D, Young T, Del Toro K, Chan V, Ianiro C, Hunt K, Kleinmahon J. Acute fulminant myocarditis in a pediatric patient with COVID-19 infection. Pediatrics August 1, 2020;(2):146.
[40] Bailey AL, Dmytrenko O, Greenberg L, Bredemeyer AL, Ma P, Liu J, Penna V, Winkler ES, Sviben S, Brooks E, Nair AP. SARS-CoV-2 infects human engineered heart tissues and models COVID-19 myocarditis. Basic to Translational Science April 1, 2021;6(4):331−45.
[41] Marshall M, Ferguson ID, Lewis P, Jaggi P, Gagliardo C, Collins JS, Shaughnessy R, Caron R, Fuss C, Corbin KJ, Emuren L. Symptomatic acute myocarditis in 7 adolescents after Pfizer-BioNTech COVID-19 vaccination. Pediatrics September 1, 2021;(3):148.
[42] Chin SE, Bhavsar SM, Corson A, Ghersin ZJ, Kim HS. Cardiac complications associated with COVID-19, MIS-C, and mRNA COVID-19 vaccination. Pediatr Cardiol March 2022;8:1−6.
[43] Huynh A, Sehn JK, Goldfarb IT, Watkins J, Torous V, Heerema-McKenney A, Roberts DJ. SARS-CoV-2 placentitis and intraparenchymal thrombohematomas among COVID-19 infections in pregnancy. JAMA Netw Open March 1, 2022;5(3):e225345.
[44] Ferrara A, Hedderson MM, Zhu Y, Avalos LA, Kuzniewicz MW, Myers LC, Ngo AL, Gunderson EP, Ritchie JL, Quesenberry CP, Greenberg M. Perinatal complications in individuals in California with or without SARS-CoV-2 infection during pregnancy. JAMA Intern Med May 1, 2022;182(5):503−12.
[45] Reagan-Steiner S, Bhatnagar J, Martines RB, Milligan NS, Gisondo C, Williams FB, Lee E, Estetter L, Bullock H, Goldsmith CS, Fair P. Detection of SARS-CoV-2 in neonatal autopsy tissues and placenta. Emerg Infect Dis March 2022;28(3):510.
[46] Watkins JC, Torous VF, Roberts DJ. Defining severe acute respiratory syndrome coronavirus 2 (SARS-CoV-2) placentitis: a report of 7 cases with confirmatory in situ hybridization, distinct histomorphologic features, and evidence of complement deposition. Arch Pathol Lab Med November 2021;145(11):1341−9.
[47] Shanes ED, Mithal LB, Otero S, Azad HA, Miller ES, Goldstein JA. Placental pathology in COVID-19. Am J Clin Pathol June 8, 2020;154(1):23−32.
[48] Colmenero I, Santonja C, Alonso-Riaño M, Noguera-Morel L, Hernández-Martín A, Andina D, Wiesner T, Rodríguez-Peralto JL, Requena L, Torrelo A. SARS-CoV-2 endothelial infection causes COVID-19 chilblains: histopathological, immunohistochemical and ultrastructural study of seven paediatric cases. Br J Dermatol October 2020;183(4):729−37.
[49] Torrelo A, Andina D, Santonja C, Noguera-Morel L, Bascuas-Arribas M, Gaitero-Tristán J, Alonso-Cadenas JA, Escalada-Pellitero S, Á H-M, de la Torre-Espi M, Colmenero I.

Erythema multiforme-like lesions in children and COVID-19. Pediatr Dermatol May 2020;37(3):442–6.

[50] Suchonwanit P, Leerunyakul K, Kositkuljorn C. Cutaneous manifestations in COVID-19: lessons learned from current evidence. J Am Acad Dermatol July 1, 2020;83(1):e57–60.

[51] Sadeghzadeh-Bazargan A, Rezai M, Najar Nobari N, Mozafarpoor S, Goodarzi A. Skin manifestations as potential symptoms of diffuse vascular injury in critical COVID-19 patients. J Cutan Pathol October 2021;48(10):1266–76.

[52] Mascitti H, Jourdain P, Bleibtreu A, Jaulmes L, Dechartres A, Lescure X, Yordanov Y, Dinh A. Prognosis of rash and chilblain-like lesions among outpatients with COVID-19: a large cohort study. Eur J Clin Microbiol Infect Dis October 2021;40(10):2243–8.

[53] Gehlhausen JR, Little AJ, Ko CJ, Emmenegger M, Lucas C, Wong P, Klein J, Lu P, Mao T, Jaycox J, Wang E. Lack of association between pandemic chilblains and SARS-CoV-2 infection. Proc Natl Acad Sci USA March 1, 2022;119(9). e2122090119.

[54] Solomon IH, Normandin E, Bhattacharyya S, Mukerji SS, Keller K, Ali AS, Adams G, Hornick JL, Padera Jr RF, Sabeti P. Neuropathological features of covid-19. N Engl J Med September 3, 2020;383(10):989–92.

[55] McCray Jr PB, Pewe L, Wohlford-Lenane C, Hickey M, Manzel L, Shi L, Netland J, Jia HP, Halabi C, Sigmund CD, Meyerholz DK. Lethal infection of K18-hACE2 mice infected with severe acute respiratory syndrome coronavirus. J Virol January 15, 2007;81(2):813–21.

[56] Desforges M, Coupanec AL, Brison É, Meessen-Pinard M, Talbot PJ. Neuroinvasive and neurotropic human respiratory coronaviruses: potential neurovirulent agents in humans. Infectious Diseases and Nanomedicine I 2014:75–96.

[57] Gutierrez Amezcua JM, Jain R, Kleinman G, Muh CR, Guzzetta M, Folkerth R, Snuderl M, Placantonakis DG, Galetta SL, Hochman S, Zagzag D. COVID-19-induced neurovascular injury: a case series with emphasis on pathophysiological mechanisms. SN Comprehensive Clinical Medicine November 2020;2(11):2109–25.

Chapter 3

Signs and symptoms commonly seen in COVID-19 in newborns, children, and adolescents and pediatric subjects

Giuseppina Malcangi[1,†], Alessio Danilo Inchingolo[1,†], Angelo Michele Inchingolo[1,†], Luigi Santacroce[1], Grazia Marinelli[1], Antonio Mancini[1], Luigi Vimercati[1], Maria Elena Maggiore[1], Maria Teresa D'Oria[1,2], Damiano Nemore[1], Arnaldo Scardapane[1], Biagio Rapone[1], Maria Franca Coscia[3], Ioana Roxana Bordea[4], Edit Xhajanka[5], Antonio Scarano[6], Marco Farronato[7], Gianluca Martino Tartaglia[7], Delia Giovanniello[8], Ludovica Nucci[9], Rosario Serpico[9], Mariantonietta Francavilla[10], Loredana Capozzi[11], Antonio Parisi[11], Marina Di Domenico[12], Felice Lorusso[6], Maria Contaldo[9,‡], Francesco Inchingolo[1,‡] and Gianna Dipalma[1,‡]

[1]*Department of Interdisciplinary Medicine, University of Medicine Aldo Moro, Bari, Italy;* [2]*Department of Medical and Biological Sciences, University of Udine, Udine, Italy;* [3]*Department of Basical Medical Sciences, Neurosciences and Sense Organs (S.M.B.N.O.S.), University of Medicine Aldo Moro, Bari, Italy;* [4]*Department of Oral Rehabilitation, Faculty of Dentistry, Iuliu Hațieganu University of Medicine and Pharmacy, Cluj-Napoca, Romania;* [5]*Department of Dental Prosthesis, Medical University of Tirana, Rruga e Dibrës, U.M.T., Tirana, Albania;* [6]*Department of Innovative Technologies in Medicine and Dentistry, University of Chieti-Pescara, Chieti, Italy;* [7]*UOC Maxillo-facial Surgery and Dentistry, Department of Biomedical, Surgical and Dental Sciences, School of Dentistry, Fondazione IRCCS Ca Granda, Ospedale Maggiore Policlinico, University of Milan, Milan, Italy;* [8]*Department of Toracic Surgery, Hospital "San Camillo Forlanini", Rome, Italy;* [9]*Multidisciplinary Department of Medical-Surgical and Dental Specialties, University of Campania Luigi Vanvitelli, Naples, Italy;* [10]*Unit of Pediatric Imaging, Giovanni XXIII Hospital, Bari;* [11]*Istituto Zooprofilattico Sperimentale Della Puglia e della Basilicata, Foggia, Italy;* [12]*Department of Precision Medicine, University of Campania Luigi Vanvitelli, Naples, Italy*

†. These authors contributed equally to this work as co-first authors.
‡. These authors contributed equally to this work as co-last authors.

Clinical Management of Pediatric COVID-19. https://doi.org/10.1016/B978-0-323-95059-6.00007-3

Background

Since the first moment of the appearance of the severe acute respiratory syndrome coronavirus 2 (SARS-CoV-2) in November 2019, the clinical trend of the pandemic, declared on March 11, 2020, immediately became more serious and critical in the adult population [1–5]. The percentage of the pediatric population hospitalized for coronavirus disease 2019 (COVID-19) appeared immediately lower than that of adults although there were also serious outcomes in children and adolescents [4,6–10]. Adults and children have experienced similar symptoms of COVID-19, except that children's symptoms are usually milder and similar to colds [11–17]. Most of them recover within one to 2 weeks. The most common symptoms are: fever, cough that becomes productive, sore throat, anosmia, ageusia, dysgeusia, gastrointestinal symptoms, such as nausea, vomiting, abdominal pain, or diarrhea, skin changes, such as discolored areas, on the feet and hands, asthenia, chills, body aches, severe headache, nasal congestion [18,19]. The presence of obesity, diabetes, asthma, and comorbidities, such as congenital heart disease, genetic diseases, or situations affecting the nervous system or metabolism, predispose to greater risk of serious illnesses with COVID-19 [3,19,20]. Studies also report COVID-19 rates in black Hispanic and non-Hispanic children compared to white non-Hispanic children [19,21]. Scientific data reported in the Pediatric Intensive Care Unit (PICU) on COVID-19 are insufficient. Specific data on school transmission of SARS-CoV-2 from the first wave are lacking, due to their early closure at the start of the pandemic in many countries [22,23].

Recent studies on the Italian population also report a percentage of 1.8% of confirmed infected cases, with a slight prevalence in males and an average age of 11 years. The hospitalizations registered are 13%, and of these, 3.5% in intensive care units. The risk increases in neonates and in the presence of comorbidities [24]. Necessary was the strategic obligation to close schools all over the world to control the spread of SARS-CoV-2. In the first wave, the peak was at the end of March, in the second wave, at the end of August and first half of September. In the second wave, in all age groups, there was a reduction of the infection. In the transition period (May–September 2020), adolescents (13–17) became more infected (41.3%) than children (7–12 years, 28%; 2–6 years, 21.0%) and to infants (0–1 year, 9.7%). Only 8% of this required hospitalization, and of these children, less than 1-year-old required intensive care. Infected pediatric cases detected during the first and second epidemic waves were similar to the lockdown period but less severe than children (2%). 7.2% represented by infants under 1 year [25,26]. All studies have found that, with the reopening of schools, the lack of attention to compliance with prevention rules, such as hand washing, physical distancing, and the use of masks has led to the emergence of new outbreaks among students and personal through the most recent viral vector variants (Fig. 3.1) [27–35]. The data show a greater diffusion in secondary schools than in

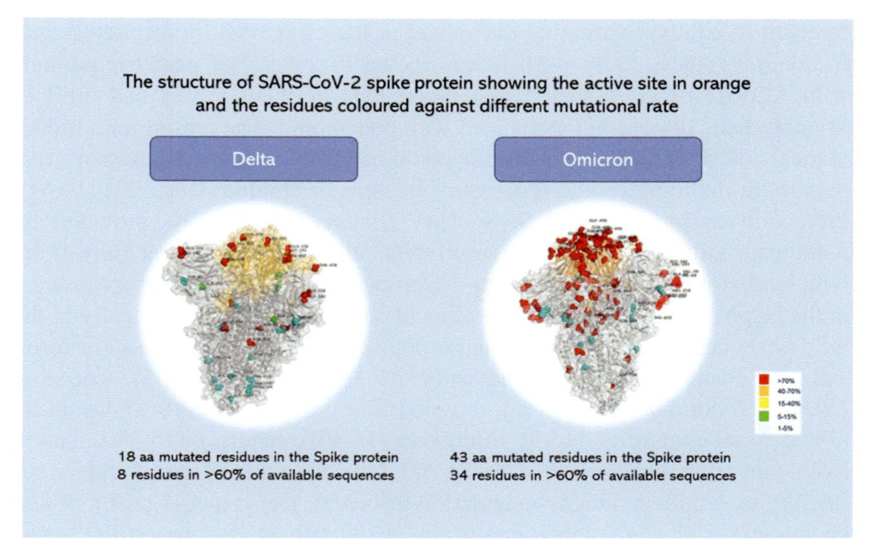

FIGURE 3.1 Spike protein mutations in Delta and Omicron variants of severe acute respiratory syndrome coronavirus 2 (SARS-CoV 2).

kindergartens and where secondary transmission is almost absent even among school staff [32]. The disadvantaged social background of children increases the risk of infection with the SARS-CoV-2 virus [34]. Since the pediatric population becomes asymptomatic or paucisymptomatic compared to adults and since only symptomatic subjects were tested during the first epidemic phase, pediatric cases, even if sporadic, were not detected, leaving unreliable overall data. After the first lockdown and during the second wave, the clearer diagnostic protocols associated with digital tracking have revealed a greater number of cases, including asymptomatic ones [36–38]. In a study conducted in the United States of hospitalizations of children and adolescents from March 2020 to August 2021, it was found that weekly hospitalization rates due to COVID-19 had increased rapidly in the population aged 0–17 between the end of June and mid-August 2021, and the percentage was almost 10 times higher than the rate 7 weeks earlier among children aged 0–4 [39]. This trend coincides with the diffusion of the Delta variant, which immediately proved to be highly transmissible.

Immunological pathogenesis and COVID-19 clinical manifestations in children

Studies report that symptomatic pediatric cases of COVID-19 have a greater persistence of SARS-CoV-2 in the upper respiratory tract and feces, manifesting greater secretion in the other respiratory tracts and gastroenteritis (less frequent in adults) facilitating the spread of the virus both by respiratory and fecal–oral routes [19,40–43]. It was shown in a study of pediatric patients with COVID-19 (1–6 years) that about half of them (53%) had mild or asymptomatic disease but associated with pneumonia. The reason for a milder clinical course of COVID-19 disease could be linked to some mechanisms that interact in the immune and respiratory systems of children (Fig. 3.2) [19,44]. Protection during pediatric SARS-CoV-2 infection could be provided by pulmonary infiltrates. This is because lymphocytes stimulate the activity of the lymphoid structure associated with the bronchi after an inflammatory process in the respiratory system [45]. The clinical course that occurs frequently in the mild or moderate form in the pediatric population is similar to the adult form, and it is characterized by low fever (43.1%), dry cough (43.4%), wheezing (20.4%), abdominal pain (0.4%), fatigue (5.1%), muscle pain (6%), headache (38%), nasal congestion (15.3), rhinorrhea (16.4%), sneezing (15.3%), nausea (6%), and odynophagia (8%) [46,47]. While in the severe form, dyspnea, refractory metabolic acidosis, coagulation disorders, tachypnea (12.6%), severe pneumonia, and septic shock can occur [46,48,49]. Studies report the emergence of a new and rare syndromes (2 out of 100,000 people under the age of 21) (Fig. 3.3): acute multisystem inflammatory syndrome (MIS-C) and Kawasaki syndrome. Occurs after 2–4 weeks with high fever (≥38°C), heart and/or gastrointestinal problems, encephalopathy, elevated inflammatory markers, rash, changes in the hands and feet with red palms and feet, conjunctivitis bilateral nonpurulent, signs of inflammation of the cutaneous mucus, with

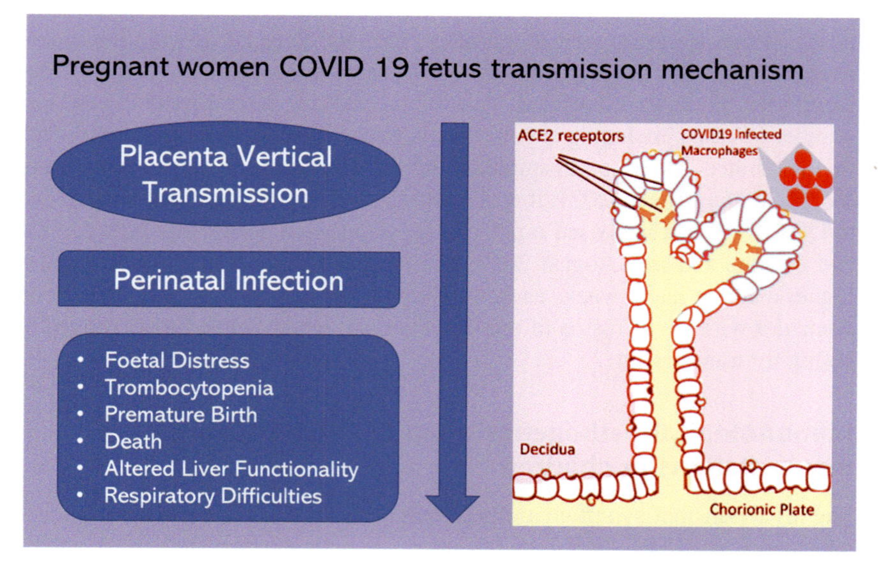

FIGURE 3.2 Transmission mechanisms of the severe acute respiratory syndrome coronavirus 2 (SARS-CoV-2) in pregnant subject and vertical alterations to the fetus.

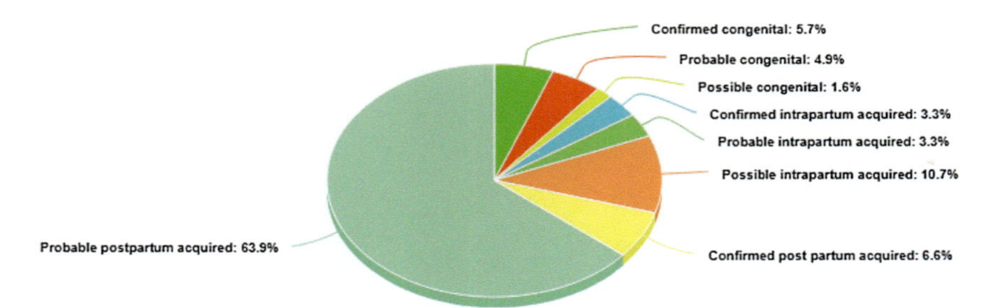

FIGURE 3.3 Distribution and frequency of the perinatal adverse effect produced by the severe acute respiratory syndrome coronavirus 2 (SARS-CoV 2).

redness of the lips and oral mucosa (with or without hard edema, skin rash, and unilateral enlargement of the cervical lymph nodes [50–52]. There is a general systemic involvement even if not all symptoms present at the same time. MIS-C in COVID-19 onset in adolescence with prevalence of abdominal symptoms and left ventricular systolic dysfunction [50,53–58].

Clinical and radiological diagnostic protocols and findings

In one study, 20 hospitalized pediatric patients with COVID-19 infection confirmed by PCR throat swab test reporting clinical, laboratory, and CT characteristics of the chest during January 23 and February 8, 2020. As the clinical situation was not serious for most of the adolescent patients, non-contrast CT was performed, as simple chest radiography was not able to identify all lung lesions [5,59,60]. Subpleural lesions with localized inflammatory infiltration were present in all children. Thirty percent had unilateral lung lesions, 50% bilateral lung lesions, and 20% had no chest CT abnormality. 60% had ground glass opacity, 20% fine mesh shade, and finally, 15% had small nodules. 65% had had close contact with family members with a certain diagnosis of COVID-19 infection (Figs. 3.4, 3.5 and 3.6) [61–66]. Cough and fever were the most frequent symptoms, respectively, 65% and 60%, whereas diarrhea and rhinorrhea in a lower percentage. Laboratory data reported that creatine kinase-MB increased in 75% of cases, C-reactive protein (CRP) increased in 45%, and in 80%, there was an increase in procalcitonin (PCT), not detectable in adults. PCT is a marker of bacterial infection that may be induced by bacteriotoxin but suppressed by interferon [67]. The procalcitonin concentration increases in bacterial infections but remains low in viral infections and inflammatory diseases [68]. The PCT value appears to have better specificity, sensitivity, and prognostic value than CPR, IL-6, and interferon alpha in children to diagnose between viral and bacterial infections. PCT is found to have a specificity for a viral infection of 94%. For the differentiation between viral and bacterial infections, CRP, IL-6, and interferon alpha have a sensitivity of 85%, 48%, and 76% and 73%, 85%, and 92%,

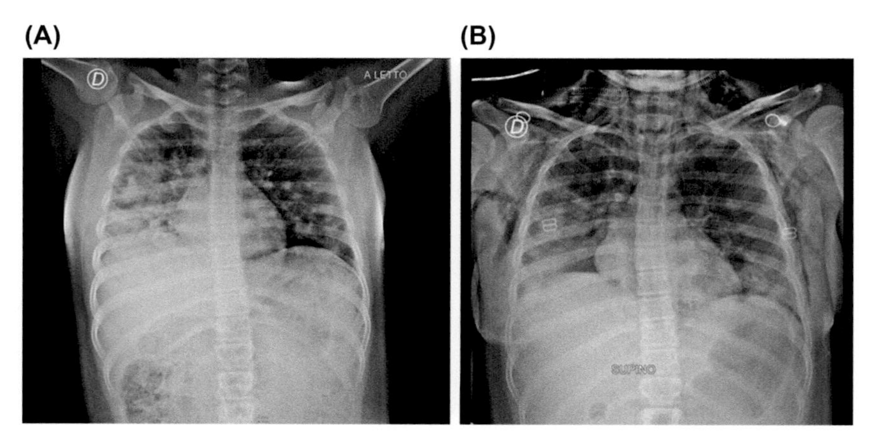

FIGURE 3.4 (A) Anteroposterior chest radiograph of a 13-year-old girl on the day of admission shows multifocal airspace consolidations, ground glass opacities, and reticular opacities in both lungs, especially in the right one. (B) anteroposterior chest radiograph of the same girl 4 days after admission showing bilateral pneumothorax, pneumomediastinum and diffuse soft tissue emphysema.

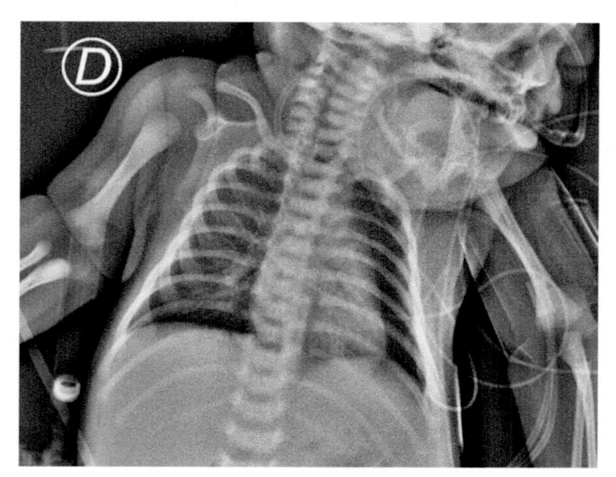

FIGURE 3.5 Anteroposterior chest radiograph of a 12-day-old boy admitted for fever and cough shows mild bilateral perihilar peribronchial wall thickening.

respectively [68]. Finding from the data that 40% of hospitalized pediatric patients were co-infected, routine antibacterial treatment in these patients should be considered. Unfortunately, CT imaging alone is not sufficient for diagnosing COVID-19 pneumonia, especially in the presence of co-infection with other pathogens. The immediacy of follow-up associated with chest CT screening, evaluating the detection of pathogens, is a recommended clinical protocol for children.

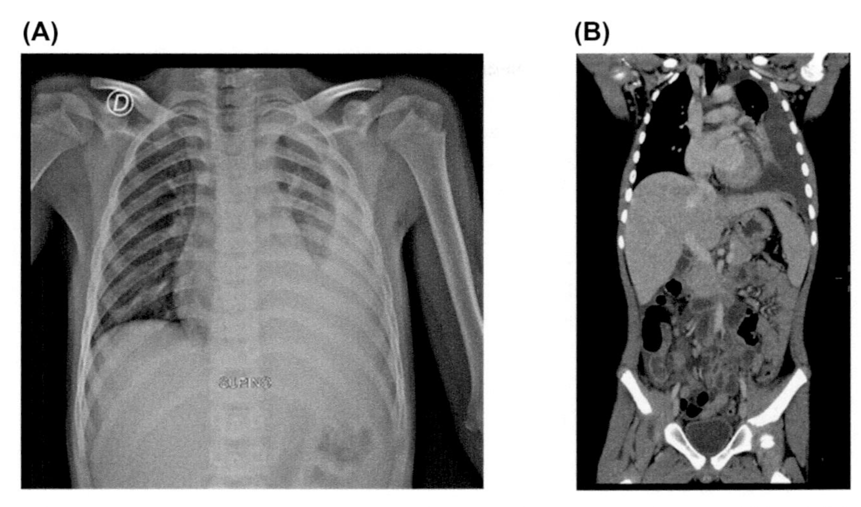

FIGURE 3.6 An 11-year-old patient with multisystem inflammatory syndrome in children (MIS-C) associated with coronavirus disease 2019 (COVID-19). (A) Anteroposterior chest radiograph displays left pleural effusion with atelectasis of the adjacent parenchyma. (B) CT chest and abdominal scan showing the same thoracic findings and parietal wall thickening of the right colon in keeping with the inflammatory syndrome.

Vaccination campaign and clinical symptoms in vaccinated children

According to COVID-NET data, the Food and Drug Administration (FDA) authorization for vaccination in the age group of 12–17 years has shown a 100% efficacy in preventing hospital admissions associated with COVID-19 in adolescents between the end of June and the end of July 2021. Hospitalization rates were 10 times higher for the unvaccinated population than for fully vaccinated adolescents [39,69,70]. In both the first wave and during the onset of the Delta variant, one in four children or adolescents required intensive care [71,72]. Emergency use of the Pfizer-BioNTech COVID-19 vaccine for this age group has been authorized by the FDA (May 11, 2021) and the European Medicines Agency (EMA) (May 28, 2021). Two injections are planned, given the same dose of Pfizer-BioNTech COVID-19 vaccine 3 weeks apart for people aged 16 and over. The second dose could be administered possibly up to 6 weeks after the first dose. Data show that this vaccine is effective in reducing symptomatic SARS-CoV-2 infections in 90.7% in children aged 12–15 years [73–76]. The US FDA (October 29, 2021) and the EMA (November 25, 2021) have also authorized the emergency use of a Pfizer-BioNTech COVID-19 vaccine for the age group between 5 and 12 years. Two injections are given, given 3 weeks apart. It contains a lower dose (1/3 of the authorized dosage for adults and adolescents) than the Pfizer-BioNTech COVID-19 vaccine used for people 12 years of age and older. Data show that this vaccine is approximately 90.7% effective in preventing symptomatic COVID-19 7 days after the second dose in participants aged 5–11 years without prior evidence of SARS-CoV-2 infection [74,76,77]. The inability to vaccinate children so far has made this age group particularly vulnerable to infections [76,78–80]. In accordance to the symptoms stratification, the vaccinated adolescent showed a more frequent mild for compared to the severe presentation, with a higher prevalence of fever, cough, headache, and nasal congestion [81,82]. Moreover, a recent study showed that the vaccinated children and adolescent affected by Omicron variant spent an average of one-half day less symptomatic convalescence period compared to unvaccinated young subjects [81].

Maternal and fetal COVID-19 symptoms

Data reported on a study that analyzed 20 placentas from mothers infected with SARS-CoV-2 show that 10 out of 20 cases had fetal vascular malperfusion or vascular thrombosis related to COVID-19-associated hypercoagulability [83]. The data issued in Nature Medicine, which analyzed the transmission patterns of COVID-19 in six countries, including Italy, reveal that children and young people under the age of 20, often asymptomatic or paucisymptomatic, become infected for a value that is about half of that of people over 20 years of age [84,85]. The data derived from a multicenter study on 82 health institutions from 25 European countries confirm that the SARS-CoV-2 infection occurs mildly in both children and newborns. The percentage for which intensive care is required is very low in both children and adolescents, and in these cases, prolonged forced respiratory assistance is required (1 week or more). Rarely, the clinical course worsens to death, which is generally rare [86]. The data show that in this population, the presence of a viral co-infection, infected with both SARS-CoV-2 and one or more viral agents, predisposes more to admission to the intensive care unit than those infected only with SARS-CoV-2 [60,86,87]. Data from a Chinese study report the percentage of 1%–2% of cases of SARS-CoV-2 infections in children, most with a favorable clinical course, even in newborns, which represent 0.6% of an evolution critical of the disease (Figs. 3.2 and 3.3), and of these 50% are less than 1 year old and with the presence of comorbidities that can cause death, albeit rare. 0.9% covers the range from 0 to 10 years and 1.2% that between 10 and 19 years [23,88,89]. The most common positivity was associated to postpartum infection (~70%), while the congenital or intrapartum SARS-CoV-2 infection was associated respectively for ~10% and ~70% (Fig. 3.2). The placenta vertical transmission was associated to fetus distress, thrombocytopenia, premature birth, death, and altered liver and respiratory function (Fig. 3.3).

Impact of COVID-19 on pediatric mental health

Particular attention was given to the psychological well-being of children and adolescents during COVID-19 due to direct and indirect causes.

In the first year of the pandemic, studies show that mental disorders in children and adolescents increased significantly, by 25.2% and 20.5%, respectively. During COVID-19, mainly older adolescents and girls showed a more secondary marked sense of anxiety, depression, uncertainty, and loneliness [90–92]. On the contrary, thanks to forced isolation, there has been a decrease in the use of drugs [90]. Emotional-behavioral disorders (anxiety, depression) predisposed the population of "pandemic adolescents" to self-destructive behaviors, such as binge drinking and uncontrolled eating habits, often associated with an obsessive use of social media [92–94]. Furthermore, reduced or absent motor activity, associated with a change in eating habits in children and adolescents, has led to an increase in obesity and overweight in this population [93,95,96]. A study on learning from home, in the ability to concentrate, in the population aged between 6 and 14 showed a difficulty in almost 30% of children. All data confirm that the difficulty of adapting to isolation in quarantine, to restrictive measures have exposed developing children to stress, who have shown difficulty falling asleep (28%) and desire to sleep with their parents [97]. There is also the difficulty in elaborating the meaning of the pandemic and in the sad experiences of family mourning. Many studies have reported the spread of "languishing" that is a state of absence of well-being, purpose, and joy that in the medium and long term could lead to dramatic consequences on the physical and mental health of children [98].

Acknowledgments

The authors declare no acknowledgment for the present research.

Patents

Author Contributions: Conceptualization, G.M. (Giuseppina Malcangi), F.I., G.D., F.L., A.S. (Antonio Scarano), A.S. (Arnaldo Scardapane) and I.R.B.; methodology, G.M. (Giuseppina Malcangi), A.M.I., A.D.I., G.D., D.N., M.F.C., M.F., and G.M. (Grazia Marinelli); software, A.M., A.D.I, and G.M. (Giuseppina Malcangi), A.S. (Antonio Scarano), A.S. (Arnaldo Scardapane), F.L., I.R.B., E.X., and F.L.; validation, L.S., A.M., L.V., M.E.M., M.T.D., L.C., A.P., B.R., M.F., G.D., F.I., I.R.B., E.X., M.F., and G.M.T.; formal analysis, I.R.B., G.D., F.I., and A.S. (Antonio Scarano), F.L., A.M.I., and A.D.I.; investigation, M.F., D.N., M.F.C., G.D., and I.R.B.; resources, A.D.I., D.G., D.M., A.M.I., E.X., and G.M. (Grazia Marinelli); data curation, G.M. (Giuseppina Malcangi), M.F.C., L.S., and A.S. (Antonio Scarano), A.S. (Arnaldo Scardapane); writing—original draft preparation, F.L., G.M. (Giuseppina Malcangi), F.I., and G.D. writing—review and editing, F.I, A.S. (Antonio Scarano), F.L., I.R.B., and A.D.I.; visualization, G.M. (Giuseppina Malcangi), F.L. F.I., and I.R.B.; supervision, F.I, F.L. and A.S. (Antonio Scarano), I.R.B., G.D., R.S., M.C., G.M.T., M.F., and A.D.I.; project administration, F.I., G.D.; funding acquisition, F.I., A.M., M.E.M., and A.D.I.

Funding: This research received no external funding.

Statement: The present prospective clinical study was based in University of Bari (Italy), in full accordance with ethical principles, including the World Medical Association Declaration of Helsinki and the additional requirements of Italian law. Furthermore, the University of Bari, Italy, classified the study to be exempt from ethical review as it carries only negligible risk and involves the use of existing data that contains only nonidentifiable data about human beings. The patient signed a written informed consent form.
Informed Consent Statement: Informed consent was obtained from the subjects involved in the study. Written informed consent has been obtained from the patient to publish this article.

Data Availability Statement: All experimental data to support the findings of this study are available contacting the corresponding author upon request. The authors have annotated the entire data building process and empirical techniques presented in the paper.

Conflicts of Interest: The authors declare no conflict of interest.

References

[1] Baronio M, Freni-Sterrantino A, Pinelli M, Natalini G, Tonini G, Marri M, Baglivo M, Sabatini T, Maltese PE, Chiurazzi P, Michelini S, Morreale G, Ascione A, Notaro P, Bertelli M. Italian SARS-CoV-2 patients in intensive care: towards an identikit for subjects at risk? Eur Rev Med Pharmacol Sci 2020;24:9698−704. https://doi.org/10.26355/eurrev_202009_23061.

[2] Hoehl S, Rabenau H, Berger A, Kortenbusch M, Cinatl J, Bojkova D, Behrens P, Böddinghaus B, Götsch U, Naujoks F, Neumann P, Schork J, Tiarks-Jungk P, Walczok A, Eickmann M, Vehreschild MJGT, Kann G, Wolf T, Gottschalk R, Ciesek S. Evidence of SARS-CoV-2 infection in returning travelers from Wuhan, China. N Engl J Med 2020;382:1278−80. https://doi.org/10.1056/NEJMc2001899.

[3] Jutzeler CR, Bourguignon L, Weis CV, Tong B, Wong C, Rieck B, Pargger H, Tschudin-Sutter S, Egli A, Borgwardt K, Walter M. Comorbidities, clinical signs and symptoms, laboratory findings, imaging features, treatment strategies, and outcomes in adult and pediatric patients with COVID-19: a systematic review and meta-analysis. Trav Med Infect Dis 2020;37:101825. https://doi.org/10.1016/j.tmaid.2020.101825.

[4] Li Q, Guan X, Wu P, Wang X, Zhou L, Tong Y, Ren R, Leung KSM, Lau EHY, Wong JY, Xing X, Xiang N, Wu Y, Li C, Chen Q, Li D, Liu T, Zhao J, Liu M, Tu W, Chen C, Jin L, Yang R, Wang Q, Zhou S, Wang R, Liu H, Luo Y, Liu Y, Shao G, Li H, Tao Z, Yang Y, Deng Z, Liu B, Ma Z, Zhang Y, Shi G, Lam TTY, Wu JT, Gao GF, Cowling BJ, Yang B, Leung GM, Feng Z. Early transmission dynamics in Wuhan, China, of novel coronavirus-infected pneumonia. N Engl J Med 2020;382:1199−207. https://doi.org/10.1056/NEJMoa2001316.

[5] Xia W, Shao J, Guo Y, Peng X, Li Z, Hu D. Clinical and CT features in pediatric patients with COVID-19 infection: different points from adults. Pediatr Pulmonol 2020;55:1169−74. https://doi.org/10.1002/ppul.24718.

[6] Bellocchio L, Bordea IR, Ballini A, Lorusso F, Hazballa D, Isacco CG, Malcangi G, Inchingolo AD, Dipalma G, Inchingolo F, Piscitelli P, Logroscino G, Miani A. Environmental issues and neurological manifestations associated with COVID-19 pandemic: new aspects of the disease? Int J Environ Res Publ Health 2020;17:8049. https://doi.org/10.3390/ijerph17218049.

[7] Bordea IR, Xhajanka E, Candrea S, Bran S, Onișor F, Inchingolo AD, Malcangi G, Pham VH, Inchingolo AM, Scarano A, Lorusso F, Isacco CG, Aityan SK, Ballini A,

Dipalma G, Inchingolo F. Coronavirus (SARS-CoV-2) pandemic: future challenges for dental practitioners. Microorganisms 2020;8:1704. https://doi.org/10.3390/microorganisms8111704.

[8] Delahoy MJ, Ujamaa D, Whitaker M, O'Halloran A, Anglin O, Burns E, Cummings C, Holstein R, Kambhampati AK, Milucky J, Patel K, Pham H, Taylor CA, Chai SJ, Reingold A, Alden NB, Kawasaki B, Meek J, Yousey-Hindes K, Anderson EJ, Openo KP, Teno K, Weigel A, Kim S, Leegwater L, Bye E, Como-Sabetti K, Ropp S, Rudin D, Muse A, Spina N, Bennett NM, Popham K, Billing LM, Shiltz E, Sutton M, Thomas A, Schaffner W, Talbot HK, Crossland MT, McCaffrey K, Hall AJ, Fry AM, McMorrow M, Reed C, Garg S, Havers FP, Kirley PD, McLafferty S, Armistead I, Fawcett E, Ward K, Lynfield R, Danila R, Khanlian S, Angeles K, Engesser K, Rowe A, Felsen C, Bushey S, Abdullah N, West N, Markus T, Hill M, George A. Hospitalizations associated with COVID-19 among children and adolescents—COVID-NET, 14 states, March 1, 2020–August 14, 2021. MMWR Morb Mortal Wkly Rep 2021;70:1255–60. https://doi.org/10.15585/mmwr.mm7036e2.

[9] Inchingolo AD, Dipalma G, Inchingolo AM, Malcangi G, Santacroce L, D'Oria MT, Isacco CG, Bordea IR, Candrea S, Scarano A, Morandi B, Del Fabbro M, Farronato M, Tartaglia GM, Balzanelli MG, Ballini A, Nucci L, Lorusso F, Taschieri S, Inchingolo F. The 15-months clinical experience of SARS-CoV-2: a literature review of therapies and adjuvants. Antioxidants 2021;10:881. https://doi.org/10.3390/antiox10060881.

[10] Scarano A, Inchingolo F, Lorusso F. Environmental disinfection of a dental clinic during the covid-19 pandemic: a narrative insight. BioMed Res Int 2020;8896812:1–15.

[11] Balzanelli MG, Distratis P, Catucci O, Cefalo A, Lazzaro R, Inchingolo F, et al. Mesenchymal stem cells: the secret children's weapons against the SARS-CoV-2 lethal infection. Appl Sci 2021;11:1696–702.

[12] Balzanelli MG, Distratis P, Aityan SK, Amatulli F, Catucci O, Cefalo A, et al. An alternative "Trojan Horse" hypothesis for COVID-19: immune deficiency of IL-10 and SARS-CoV-2 biology. Endocr Metab Immune Disord: Drug Targets 2021;1:1–5. https://doi.org/10.2174/1871530321666210127141945.

[13] Charitos IA, Ballini A, Bottalico L, Cantore S, Passarelli PC, Inchingolo F, et al. Special features of SARS-CoV-2 in daily practice. WJCC 2020;8(18):3920–33. https://doi.org/10.12998/wjcc.v8.i18.3920.

[14] Charitos, I.A., Del Prete, R., Inchingolo, F., Mosca, A., Carretta, D., Ballini, A., Santacroce, L., n.d. What we have learned for the future about COVID-19 and healthcare management of it?.

[15] Santacroce L, Inchingolo F, Topi S, Del Prete R, Di Cosola M, Charitos IA, Montagnani M. Potential beneficial role of probiotics on the outcome of COVID-19 patients: an evolving perspective. Diabetes Metabol Syndr: Clin Res Rev 2021;15:295–301. https://doi.org/10.1016/j.dsx.2020.12.040.

[16] Santacroce L, Charitos IA, Ballini A, Inchingolo F, Luperto P, De Nitto E, Topi S. The human respiratory system and its microbiome at a glimpse. Biology 2020;9:318. https://doi.org/10.3390/biology9100318.

[17] Scarano A, Inchingolo F, Rapone B, Festa F, Tari SR, Lorusso F. Protective face masks: effect on the oxygenation and heart rate status of oral surgeons during surgery. Int J Environ Res Publ Health 2021;18:1–11.

[18] Inchingolo AD, Inchingolo AM, Bordea IR, Malcangi G, Xhajanka E, Scarano A, Lorusso F, Farronato M, Tartaglia GM, Isacco CG, Marinelli GD, Hazballa DMT, Santacroce L, Ballini A, Contaldo M, Inchingolo F, Dipalma G. Sars-cov-2 disease through

viral genomic and receptor implications: an overview of diagnostic and immunology breakthroughs. Microorganisms 2021;9. https://doi.org/10.3390/microorganisms9040793.

[19] Malcangi G, Inchingolo AD, Inchingolo AM, Santacroce L, Marinelli G, Mancini A, Vimercati L, Maggiore ME, D'Oria MT, Hazballa D, Bordea IR, Xhajanka E, Scarano A, Farronato M, Tartaglia GM, Giovanniello D, Nucci L, Serpico R, Sammartino G, Capozzi L, Parisi A, Di Domenico M, Lorusso F, Contaldo M, Inchingolo F, Dipalma G. COVID-19 infection in children, infants and pregnant subjects: an overview of recent insights and therapies. Microorganisms 2021;9:1964. https://doi.org/10.3390/microorganisms9091964.

[20] Vimercati L, De Maria L, Quarato M, Caputi A, Gesualdo L, Migliore G, Cavone D, Sponselli S, Pipoli A, Inchingolo F, Scarano A, Lorusso F, Stefanizzi P, Tafuri S. Association between long COVID and overweight/obesity. J Clin Med 2021;10.

[21] Balzanelli MG, Distratis P, Lazzaro R, Cefalo A, Catucci O, Aityan SK, Dipalma G, Vimercati L, Inchingolo AD, Maggiore ME, Mancini A, Santacroce L, Gesualdo L, Pham VH, Iacobone D, Contaldo M, Serpico R, Scarano A, Lorusso F, Toai TC, Tafuri S, Migliore G, Inchingolo AM, Nguyen KCD, Inchingolo F, Tomassone D, Gargiulo Isacco C. The vitamin D, IL-6 and the eGFR markers a possible way to elucidate the lung-heart-kidney cross-talk in COVID-19 disease: a foregone conclusion. Microorganisms 2021;9:1903. https://doi.org/10.3390/microorganisms9091903.

[22] Balzanelli MG, Distratis P, Dipalma G, Vimercati L, Inchingolo AD, Lazzaro R, Aityan SK, Maggiore ME, Mancini A, Laforgia R, Pezzolla A, Tomassone D, Pham VH, Iacobone D, Castrignano A, Scarano A, Lorusso F, Tafuri S, Migliore G, Inchingolo AM, Nguyen KCD, Toai TC, Inchingolo F, Isacco CG. Sars-CoV-2 virus infection may interfere CD34+ hematopoietic stem cells and megakaryocyte-erythroid progenitors differentiation contributing to platelet defection towards insurgence of thrombocytopenia and thrombophilia. Microorganisms 2021;9:1632. https://doi.org/10.3390/microorganisms9081632.

[23] Lee P-I, Hu Y-L, Chen P-Y, Huang Y-C, Hsueh P-R. Are children less susceptible to COVID-19? J Microbiol Immunol Infect 2020;53:371−2. https://doi.org/10.1016/j.jmii.2020.02.011.

[24] Bellino S, Punzo O, Rota MC, Del Manso M, Urdiales AM, Andrianou X, et al. COVID-19 disease severity risk factors for pediatric patients in Italy. e2020009399 Pediatrics October 2020;146(4). https://doi.org/10.1542/peds.2020-009399. Epub 2020 Jul 14. PMID: 32665373. SIN, n.d.

[25] Di Giallonardo F, Duchene S, Puglia I, Curini V, Profeta F, Cammà C, Marcacci M, Calistri P, Holmes EC, Lorusso A. Genomic epidemiology of the first wave of SARS-CoV-2 in Italy. Viruses 2020;12:E1438. https://doi.org/10.3390/v12121438.

[26] Giudice A, Barone S, Muraca D, Averta F, Diodati F, Antonelli A, Fortunato L. Can teledentistry improve the monitoring of patients during the covid-19 dissemination? A descriptive pilot study. Int J Environ Res Publ Health 2020;17:E3399. https://doi.org/10.3390/ijerph17103399.

[27] Auger KA, Shah SS, Richardson T, Hartley D, Hall M, Warniment A, Timmons K, Bosse D, Ferris SA, Brady PW, Schondelmeyer AC, Thomson JE. Association between statewide school closure and COVID-19 incidence and mortality in the US. JAMA 2020;324:859−70. https://doi.org/10.1001/jama.2020.14348.

[28] Bayham J, Fenichel EP. Impact of school closures for COVID-19 on the US health-care workforce and net mortality: a modelling study. Lancet Public Health 2020;5:e271−8. https://doi.org/10.1016/S2468-2667(20)30082-7.

[29] Domenico, L.D., Pullano, G., Sabbatini, C.E., Boëlle, P.-Y., Colizza, V., n.d. Can we safely reopen schools during COVID-19 epidemic? 21.

[30] Dub T, Erra E, Hagberg L, Sarvikivi E, Virta C, Jarvinen A, et al. Transmission of SARS-CoV-2 following exposure in school settings: experience from two Helsinki area exposure incidents. (preprint). Infectious Diseases (except HIV/AIDS) 2020;1:1–2. https://doi.org/10.1101/2020.07.20.20156018.

[31] Head JR, Andrejko KL, Cheng Q, Collender PA, Phillips S, Boser A, Heaney AK, Hoover CM, Wu SL, Northrup GR, Click K. The effect of school closures and reopening strategies on COVID-19 infection dynamics in the San Francisco Bay Area: a cross-sectional survey and modeling analysis. https://doi.org/10.1101/2020.08.06.20169797; 2020.

[32] La Rosa G, Mancini P, Bonanno Ferraro G, Veneri C, Iaconelli M, Lucentini L, Bonadonna L, Brusaferro S, Brandtner D, Fasanella A, Pace L, Parisi A, Galante D, Suffredini E. Rapid screening for SARS-CoV-2 variants of concern in clinical and environmental samples using nested RT-PCR assays targeting key mutations of the spike protein. Water Res 2021;197:117104. https://doi.org/10.1016/j.watres.2021.117104.

[33] Lee B, Hanley JP, Nowak S, Bates JHT, Hébert-Dufresne L. Modeling the impact of school reopening on SARS-CoV-2 transmission using contact structure data from Shanghai. BMC Publ Health 2020;20:1713. https://doi.org/10.1186/s12889-020-09799-8.

[34] Otte im Kampe E, Lehfeld A-S, Buda S, Buchholz U, Haas W. Surveillance of COVID-19 school outbreaks, Germany, March to August 2020. Euro Surveill 2020;25. https://doi.org/10.2807/1560-7917.ES.2020.25.38.2001645.

[35] Scarano A, Inchingolo F, Lorusso F. Facial skin temperature and discomfort when wearing protective face masks: thermal infrared imaging evaluation and hands moving the mask. Int J Environ Res Publ Health 2020;17. https://doi.org/10.3390/ijerph17134624.

[36] Kim L, Whitaker M, O'Halloran A, Kambhampati A, Chai SJ, Reingold A, Armistead I, Kawasaki B, Meek J, Yousey-Hindes K, Anderson EJ, Openo KP, Weigel A, Ryan P, Monroe ML, Fox K, Kim S, Lynfield R, Bye E, Shrum Davis S, Smelser C, Barney G, Spina NL, Bennett NM, Felsen CB, Billing LM, Shiltz J, Sutton M, West N, Talbot HK, Schaffner W, Risk I, Price A, Brammer L, Fry AM, Hall AJ, Langley GE, Garg S, Surveillance Team COVID-NET. Hospitalization rates and characteristics of children aged <18 years hospitalized with laboratory-confirmed COVID-19—COVID-NET, 14 states, March 1–July 25, 2020. MMWR Morb Mortal Wkly Rep 2020;69:1081–8. https://doi.org/10.15585/mmwr.mm6932e3.

[37] Patano A, Cirulli N, Beretta M, Plantamura P, Inchingolo AD, Inchingolo AM, Bordea IR, Malcangi G, Marinelli G, Scarano A, Lorusso F, Inchingolo F, Dipalma G. Education technology in orthodontics and paediatric dentistry during the covid-19 pandemic: a systematic review. Int J Environ Res Publ Health 2021;18.

[38] Zhao LP, Lybrand TP, Gilbert PB, Hawn TR, Schiffer JT, Stamatatos L, et al. Tracking SARS-CoV-2 spike protein mutations in the United States (2020/01–2021/03) using a statistical learning strategy. bioRxiv 2021;1:1–38. https://doi.org/10.1101/2021.06.15.448495.

[39] Murthy BP, Zell E, Saelee R, Murthy N, Meng L, Meador S, Reed K, Shaw L, Gibbs-Scharf L, McNaghten AD, Patel A, Stokley S, Flores S, Yoder JS, Black CL, Harris LQ. COVID-19 vaccination coverage among adolescents aged 12-17 years—United States, December 14, 2020–July 31, 2021. MMWR Morb Mortal Wkly Rep 2021;70:1206–13. https://doi.org/10.15585/mmwr.mm7035e1.

[40] Balzanelli MG, Distratis P, Dipalma G, Vimercati L, Catucci O, Amatulli F, Cefalo A, Lazzaro R, Palazzo D, Aityan SK, Pricolo G, Prudenzano A, D'Errico P, Laforgia R, Pezzolla A, Tomassone D, Inchingolo AD, Pham VH, Iacobone D, Materi GM, Scarano A,

Lorusso F, Inchingolo F, Nguyen KCD, Isacco CG. Immunity profiling of COVID-19 infection, dynamic variations of lymphocyte subsets, a comparative analysis on four different groups. Microorganisms 2021;9:2036. https://doi.org/10.3390/microorganisms9102036.

[41] Cruz AT, Zeichner SL. COVID-19 in children: initial characterization of the pediatric disease. Pediatrics 2020;145. https://doi.org/10.1542/peds.2020-0834.

[42] Kelvin AA, Halperin S. COVID-19 in children: the link in the transmission chain. Lancet Infect Dis 2020;20:633–4. https://doi.org/10.1016/S1473-3099(20)30236-X.

[43] Oliva S, Cucchiara S, Locatelli F. Children and fecal SARS-CoV-2 shedding: just the tip of the iceberg of Italian COVID-19 outbreak? Dig Liver Dis 2020;52:1219–21. https://doi.org/10.1016/j.dld.2020.06.039.

[44] Inchingolo AD, Inchingolo AM, Bordea IR, Malcangi G, Xhajanka E, Scarano A, Lorusso F, Farronato M, Tartaglia GM, Isacco CG, Marinelli G, D'Oria MT, Hazballa D, Santacroce L, Ballini A, Contaldo M, Inchingolo F, Dipalma G. SARS-CoV-2 disease adjuvant therapies and supplements breakthrough for the infection prevention. Microorganisms 2021;9:525. https://doi.org/10.3390/microorganisms9030525.

[45] Rangel-Moreno J, Hartson L, Navarro C, Gaxiola M, Selman M, Randall TD. Inducible bronchus-associated lymphoid tissue (iBALT) in patients with pulmonary complications of rheumatoid arthritis. J Clin Invest 2006;116:3183–94. https://doi.org/10.1172/JCI28756.

[46] Radtke T, Ulyte A, Puhan MA, Kriemler S. Long-term symptoms after SARS-CoV-2 infection in children and adolescents. JAMA 2021;326(9):869–71. https://doi.org/10.1001/jama.2021.11880.

[47] Zare-Zardini H, Soltaninejad H, Ferdosian F, Hamidieh AA, Memarpoor-Yazdi M. Coronavirus disease 2019 (COVID-19) in children: prevalence, diagnosis, clinical symptoms, and treatment. Int J Gen Med 2020;13:477–82. https://doi.org/10.2147/IJGM.S262098.

[48] Assaker R, Colas A-E, Julien-Marsollier F, Bruneau B, Marsac L, Greff B, Tri N, Fait C, Brasher C, Dahmani S. Presenting symptoms of COVID-19 in children: a meta-analysis of published studies. Br J Anaesth 2020;125:e330–2. https://doi.org/10.1016/j.bja.2020.05.026.

[49] Qiu H. Clinical and epidemiological features of 36 children with coronavirus disease 2019 (COVID-19) in Zhejiang. China: An Observational Cohort Study 2020;20:8.

[50] Kuo H-C. Kawasaki-like disease among Italian children in the COVID-19 era. J Pediatr 2020;224:179–83. https://doi.org/10.1016/j.jpeds.2020.07.022.

[51] Thompson HA, Mousa A, Dighe A, Fu H, Arnedo-Pena A, Barrett P, Bellido-Blasco J, Bi Q, Caputi A, Chaw L, De Maria L, Hoffmann M, Mahapure K, Ng K, Raghuram J, Singh G, Soman B, Soriano V, Valent F, Vimercati L, Wee LE, Wong J, Ghani AC, Ferguson NM. Severe acute respiratory syndrome coronavirus 2 (SARS-CoV-2) setting-specific transmission rates: a systematic review and meta-analysis. Clinical Infectious Diseases Ciab 2021;100. https://doi.org/10.1093/cid/ciab100.

[52] Zovi A, Musazzi UM, D'Angelo C, Piacenza M, Vimercati S, Cilurzo F. Medicines shortages and the perception of healthcare professionals working in hospitals: an Italian case study. J Interprof Educ Pract 2021;25:100472.

[53] Ebina-Shibuya R, Namkoong H, Shibuya Y, Horita N. Multisystem inflammatory syndrome in children (MIS-C) with COVID-19: insights from simultaneous familial Kawasaki disease cases. Int J Infect Dis 2020;97:371–3. https://doi.org/10.1016/j.ijid.2020.06.014.

[54] Jones VG, Mills M, Suarez D, Hogan CA, Yeh D, Segal JB, Nguyen EL, Barsh GR, Maskatia S, Mathew R. COVID-19 and Kawasaki disease: novel virus and novel case. Hosp Pediatr 2020;10:537–40. https://doi.org/10.1542/hpeds.2020-0123.

[55] Levin M. Childhood multisystem inflammatory syndrome—a new challenge in the pandemic. N Engl J Med 2020;383:393–5. https://doi.org/10.1056/NEJMe2023158.

[56] Riphagen S, Gomez X, Gonzalez-Martinez C, Wilkinson N, Theocharis P. Hyper-inflammatory shock in children during COVID-19 pandemic. Lancet 2020;395:1607–8. https://doi.org/10.1016/S0140-6736(20)31094-1.

[57] Toubiana J, Poirault C, Corsia A, Bajolle F, Fourgeaud J, Angoulvant F, Debray A, Basmaci R, Salvador E, Biscardi S, Frange P, Chalumeau M, Casanova J-L, Cohen JF, Allali S. Kawasaki-like multisystem inflammatory syndrome in children during the covid-19 pandemic in Paris, France: prospective observational study. BMJ 2020;369:m2094. https://doi.org/10.1136/bmj.m2094.

[58] Verdoni L, Mazza A, Gervasoni A, Martelli L, Ruggeri M, Ciuffreda M, Bonanomi E, D'Antiga L. An outbreak of severe Kawasaki-like disease at the Italian epicentre of the SARS-CoV-2 epidemic: an observational cohort study. Lancet 2020;395:1771–8. https://doi.org/10.1016/S0140-6736(20)31103-X.

[59] Akin M, Basciftci FA. Can white spot lesions be treated effectively? Angle Orthod 2012;82:770–5. https://doi.org/10.2319/090711.578.1.

[60] Ogimi C, Englund JA, Bradford MC, Qin X, Boeckh M, Waghmare A. Characteristics and outcomes of coronavirus infection in children: the role of viral factors and an immunocompromised state. J Pediatric Infect Dis Soc 2018;8:21–8. https://doi.org/10.1093/jpids/pix093.

[61] Bavaro DF, Poliseno M, Scardapane A, Belati A, De Gennaro N, Stabile Ianora AA, Angarano G, Saracino A. Occurrence of acute pulmonary embolism in COVID-19—a case series. Int J Infect Dis 2020;98:225–6. https://doi.org/10.1016/j.ijid.2020.06.066.

[62] Bevilacqua V, Altini N, Prencipe B, Brunetti A, Villani L, Sacco A, Morelli C, Ciaccia M, Scardapane A. Lung segmentation and characterization in COVID-19 patients for assessing pulmonary thromboembolism: an approach based on deep learning and radiomics. Electronics 2021;10:2475.

[63] Giovannetti G, De Michele L, De Ceglie M, Pierucci P, Mirabile A, Vita M, Palmieri VO, Carpagnano GE, Scardapane A, D'Agostino C. Lung ultrasonography for long-term follow-up of COVID-19 survivors compared to chest CT scan. Respir Med 2021;181:106384. https://doi.org/10.1016/j.rmed.2021.106384.

[64] Morelli C, Francavilla M, Stabile Ianora AA, Cozzolino M, Gualano A, Stellacci G, Sacco A, Lorusso F, Pedote P, De Ceglie M. The multifaceted COVID-19: CT aspects of its atypical pulmonary and abdominal manifestations and complications in adults and children. A pictorial review. Microorganisms 2021;9:2037.

[65] Sardaro A, Turi B, Bardoscia L, Ferrari C, Rubini G, Calabrese A, et al. The role of multiparametric magnetic resonance in volumetric modulated arc radiation therapy planning for prostate cancer recurrence after radical prostatectomy: a pilot study. Front Oncol 2020;8(10):603994–4004. https://doi.org/10.3389/fonc.2020.603994.

[66] Scardapane A, Villani L, Bavaro DF, Passerini F, Ianora AAS, Lucarelli NM, Angarano G, Portincasa P, Palmieri VO, Saracino A. Pulmonary artery filling defects in COVID-19 patients revealed using CT pulmonary angiography: a predictable complication? BioMed Res Int 2021;2021:8851736. https://doi.org/10.1155/2021/8851736.

[67] Simon L, Gauvin F, Amre DK, Saint-Louis P, Lacroix J. Serum procalcitonin and C-reactive protein levels as markers of bacterial infection: a systematic review and meta-analysis. Clin Infect Dis 2004;39:206–17. https://doi.org/10.1086/421997.

[68] Lorrot M, Moulin F, Coste J, Ravilly S, Guérin S, Lebon P, Lacombe C, Raymond J, Bohuon C, Gendrel D. [Procalcitonin in pediatric emergencies: comparison with C-reactive

protein, interleukin-6 and interferon alpha in the differentiation between bacterial and viral infections]. Presse Med 2000;29:128–34.

[69] Attolico I, Tarantini F, Carluccio P, Schifone CP, Delia M, Gagliardi VP, et al. Serological response following BNT162b2 anti-SARS-CoV-2 mRNA vaccination in haematopoietic stem cell transplantation patients. Br J Haematol 2021;190(4):928–31. https://doi.org/10.1111/bjh.17873.

[70] Bianchi FP, Tafuri S, Migliore G, Vimercati L, Martinelli A, Lobifaro A, Diella G, Stefanizzi P, Group OBOTCRW. BNT162b2 mRNA COVID-19 vaccine effectiveness in the prevention of SARS-CoV-2 infection and symptomatic disease in five-month follow-up: a retrospective cohort study. Vaccines 2021;9:1143. https://doi.org/10.3390/vaccines9101143.

[71] Buonsenso D, Munblit D, De Rose C, Sinatti D, Ricchiuto A, Carfi A, Valentini P. Preliminary evidence on long COVID in children. Acta Paediatr 2021;110:2208–11. https://doi.org/10.1111/apa.15870.

[72] Feldstein LR, Rose EB, Horwitz SM, Collins JP, Newhams MM, Son MBF, Newburger JW, Kleinman LC, Heidemann SM, Martin AA, Singh AR, Li S, Tarquinio KM, Jaggi P, Oster ME, Zackai SP, Gillen J, Ratner AJ, Walsh RF, Fitzgerald JC, Keenaghan MA, Alharash H, Doymaz S, Clouser KN, Giuliano JS, Gupta A, Parker RM, Maddux AB, Havalad V, Ramsingh S, Bukulmez H, Bradford TT, Smith LS, Tenforde MW, Carroll CL, Riggs BJ, Gertz SJ, Daube A, Lansell A, Coronado Munoz A, Hobbs CV, Marohn KL, Halasa NB, Patel MM, Randolph AG. Multisystem inflammatory syndrome in U.S. children and adolescents. N Engl J Med 2020;383:334–46. https://doi.org/10.1056/NEJMoa2021680.

[73] Ali K, Berman G, Zhou H, Deng W, Faughnan V, Coronado-Voges M, Ding B, Dooley J, Girard B, Hillebrand W, Pajon R, Miller JM, Leav B, McPhee R. Evaluation of mRNA-1273 SARS-CoV-2 vaccine in adolescents. N Engl J Med 2021;385:2241–51. https://doi.org/10.1056/NEJMoa2109522.

[74] Balasubramanian S, Rao NM, Goenka A, Roderick M, Ramanan AV. Coronavirus disease 2019 (COVID-19) in children—what we know so far and what we do not. Indian Pediatr 2020;57:435–42. https://doi.org/10.1007/s13312-020-1819-5.

[75] Malcangi G, Inchingolo AD, Inchingolo AM, Piras F, Settanni V, Garofoli G, Palmieri G, Ceci S, Patano A, Mancini A, Vimercati L, Nemore D, Scardapane A, Rapone B, Semjonova A, D'Oria MT, Macchia L, Bordea IR, Migliore G, Scarano A, Lorusso F, Tartaglia GM, Giovanniello D, Nucci L, Maggialetti N, Parisi A, Domenico MD, Brienza N, Tafuri S, Stefanizzi P, Curatoli L, Corriero A, Contaldo M, Inchingolo F, Dipalma G. COVID-19 infection in children and infants: current status on therapies and vaccines. Children 2022;9:249. https://doi.org/10.3390/children9020249.

[76] Woodworth KR, Moulia D, Collins JP, Hadler SC, Jones JM, Reddy SC, Chamberland M, Campos-Outcalt D, Morgan RL, Brooks O, Talbot HK, Lee GM, Bell BP, Daley MF, Mbaeyi S, Dooling K, Oliver SE. The advisory committee on immunization practices' interim recommendation for use of Pfizer-BioNTech COVID-19 vaccine in children aged 5–11 years—United States, November 2021. MMWR Morb Mortal Wkly Rep 2021;70:1579–83. https://doi.org/10.15585/mmwr.mm7045e1.

[77] Lu X, Xiang Y, Du H, Wing-Kin Wong G. SARS-CoV-2 infection in children—understanding the immune responses and controlling the pandemic. Pediatr Allergy Immunol 2020;31:449–53. https://doi.org/10.1111/pai.13267.

[78] Bianchi FP, Germinario CA, Migliore G, Vimercati L, Martinelli A, Lobifaro A, et al. BNT162b2 mRNA Covid-19 vaccine effectiveness in the prevention of SARS-CoV-2

infection: a preliminary report. J Infect Dis 2021;224(3):431–4. https://doi.org/10.1093/infdis/jiab262.

[79] Boffetta P, Violante F, Durando P, De Palma G, Pira E, Vimercati L, Cristaudo A, Icardi G, Sala E, Coggiola M. Determinants of SARS-CoV-2 infection in Italian healthcare workers: a multicenter study. Sci Rep 2021;11:1–8.

[80] Spagnolo L, Vimercati L, Caputi A, Benevento M, De Maria L, Ferorelli D, Solarino B. Role and tasks of the occupational physician during the COVID-19 pandemic. Medicina 2021;57. https://doi.org/10.3390/medicina57050479.

[81] Fowlkes AL, Yoon SK, Lutrick K, Gwynn L, Burns J, Grant L, Phillips AL, Ellingson K, Ferraris MV, LeClair LB, Mathenge C, Yoo YM, Thiese MS, Gerald LB, Solle NS, Jeddy Z, Odame-Bamfo L, Mak J, Hegmann KT, Gerald JK, Ochoa JS, Berry M, Rose S, Lamberte JM, Madhivanan P, Pubillones FA, Rai RP, Dunnigan K, Jones JT, Krupp K, Edwards LJ, Bedrick EJ, Sokol BE, Lowe A, McLeland-Wieser H, Jovel KS, Fleary DE, Khan SM, Poe B, Hollister J, Lopez J, Rivers P, Beitel S, Tyner HL, Naleway AL, Olsho LEW, Caban-Martinez AJ, Burgess JL, Thompson MG, Gaglani M. Effectiveness of 2-dose BNT162b2 (Pfizer BioNTech) mRNA vaccine in preventing SARS-CoV-2 infection among children aged 5–11 years and adolescents aged 12–15 years—protect Cohort, July 2021–February 2022. MMWR Morb Mortal Wkly Rep 2022;71:422–8. https://doi.org/10.15585/mmwr.mm7111e1.

[82] Hurst JH, McCumber AW, Aquino JN, Rodriguez J, Heston SM, Lugo DJ, et al. Age-related changes in the nasopharyngeal microbiome are associated with SARS-CoV-2 infection and symptoms among children, adolescents, and young adults. Clin Infect Dis 2022;184:1–33. https://doi.org/10.1093/cid/ciac184.

[83] Ahmed M, Advani S, Moreira A, Zoretic S, Martinez J, Chorath K, Acosta S, Naqvi R, Burmeister-Morton F, Burmeister F, Tarriela A, Petershack M, Evans M, Hoang A, Rajasekaran K, Ahuja S, Moreira A. Multisystem inflammatory syndrome in children: a systematic review. EClinicalMedicine 2020;26:100527. https://doi.org/10.1016/j.eclinm.2020.100527.

[84] Davies NG, Klepac P, Liu Y, Prem K, Jit M, Eggo RM. Age-dependent effects in the transmission and control of COVID-19 epidemics. Nat Med 2020;26:1205–11. https://doi.org/10.1038/s41591-020-0962-9.

[85] Götzinger F, Santiago-García B, Noguera-Julián A, Lanaspa M, Lancella L, Carducci FIC, Gabrovska N, Velizarova S, Prunk P, Osterman V, Krivec U, Vecchio AL, Shingadia D, Soriano-Arandes A, Melendo S, Lanari M, Pierantoni L, Wagner N, L'Huillier AG, Heininger U, Ritz N, Bandi S, Krajcar N, Roglić S, Santos M, Christiaens C, Creuven M, Buonsenso D, Welch SB, Bogyi M, Brinkmann F, Tebruegge M, Pfefferle J, Zacharasiewicz A, Berger A, Berger R, Strenger V, Kohlfürst DS, Zschocke A, Bernar B, Simma B, Haberlandt E, Thir C, Biebl A, Driessche KV, Boiy T, Brusselen DV, Bael A, Debulpaep S, Schelstraete P, Pavic I, Nygaard U, Glenthoej JP, Jensen LH, Lind I, Tistsenko M, Uustalu Ü, Buchtala L, Thee S, Kobbe R, Rau C, Schwerk N, Barker M, Tsolia M, Eleftheriou I, Gavin P, Kozdoba O, Zsigmond B, Valentini P, Ivaškeviciene I, Ivaškevicius R, Vilc V, Schölvinck E, Rojahn A, Smyrnaios A, Klingenberg C, Carvalho I, Ribeiro A, Starshinova A, Solovic I, Falcón L, Neth O, Minguell L, Bustillo M, Gutiérrez-Sánchez AM, Ibáñez BG, Ripoll F, Soto B, Kötz K, Zimmermann P, Schmid H, Zucol F, Niederer A, Buettcher M, Cetin BS, Bilogortseva O, Chechenyeva V, Demirjian A, Shackley F, McFetridge L, Speirs L, Doherty C, Jones L, McMaster P, Murray C, Child F, Beuvink Y, Makwana N, Whittaker E, Williams A, Fidler K, Bernatoniene J, Song R, Oliver Z, Riordan A. COVID-19 in children and adolescents in Europe: a multinational,

multicentre cohort study. The Lancet Child Adolescent Health 2020;4:653–61. https://doi.org/10.1016/S2352-4642(20)30177-2.

[86] Calvo C, Alcolea S, Casas I, Pozo F, Iglesias M, Gonzalez-Esguevillas M, Luz García-García M. A 14-year prospective study of human coronavirus infections in hospitalized children: comparison with other respiratory viruses. Pediatr Infect Dis J 2020;39:653–7. https://doi.org/10.1097/INF.0000000000002760.

[87] Chiu W, Cheung PCH, Ng KL, Ip PLS, Sugunan VK, Luk DCK, Ma LCK, Chan BHB, Lo KL, Lai WM. Severe acute respiratory syndrome in children: experience in a regional hospital in Hong Kong. Pediatr Crit Care Med 2003;4:279–83. https://doi.org/10.1097/01.PCC.0000077079.42302.81.

[88] Vimercati L, Stefanizzi P, De Maria L, Caputi A, Cavone D, Quarato M, Gesualdo L, Lopalco PL, Migliore G, Sponselli S, Graziano G, Larocca AMV, Tafuri S. Large-scale IgM and IgG SARS-CoV-2 serological screening among healthcare workers with a low infection prevalence based on nasopharyngeal swab tests in an Italian university hospital: perspectives for public health. Environ Res 2021;195:110793. https://doi.org/10.1016/j.envres.2021.110793.

[89] Vimercati L, De Maria L, Quarato M, Caputi A, Stefanizzi P, Gesualdo L, Migliore G, Fucilli FIM, Cavone D, Delfino MC, Sponselli S, Chironna M, Tafuri S. COVID-19 hospital outbreaks: protecting healthcare workers to protect frail patients. An Italian observational cohort study. Int J Infect Dis 2021;102:532–7. https://doi.org/10.1016/j.ijid.2020.10.098.

[90] Muzi S, Sansò A, Pace CS. What's happened to Italian adolescents during the COVID-19 pandemic? A preliminary study on symptoms, problematic social media usage, and attachment: relationships and differences with pre-pandemic peers. Front Psychiatr 2021;12:590543. https://doi.org/10.3389/fpsyt.2021.590543.

[91] Racine N, McArthur BA, Cooke JE, Eirich R, Zhu J, Madigan S. Global prevalence of depressive and anxiety symptoms in children and adolescents during COVID-19: a meta-analysis. JAMA Pediatr 2021;175(11):1142–50. https://doi.org/10.1001/jamapediatrics.2021.2482.

[92] Thorisdottir IE, Asgeirsdottir BB, Kristjansson AL, Valdimarsdottir HB, Tolgyes EMJ, Sigfusson J, et al. Depressive symptoms, mental wellbeing, and substance use among adolescents before and during the COVID-19 pandemic in Iceland: a longitudinal, population-based study. Lancet Psychiatr 2021;8(8):663–72.

[93] Cipolla C, Curatola A, Ferretti S, Giugno G, Condemi C, Delogu AB, Birritella L, Lazzareschi I. Eating habits and lifestyle in children with obesity during the COVID19 lockdown: a survey in an Italian center. Acta Biomed 2021;92. https://doi.org/10.23750/abm.v92i2.10912. e2021196.

[94] Nissen JB, Højgaard DRMA, Thomsen PH. The immediate effect of COVID-19 pandemic on children and adolescents with obsessive compulsive disorder. BMC Psychiatr 2020;20:511. https://doi.org/10.1186/s12888-020-02905-5.

[95] Rundle AG, Park Y, Herbstman JB, Kinsey EW, Wang YC. COVID-19 related school closings and risk of weight gain among children. Obesity 2020;28:1008–9. https://doi.org/10.1002/oby.22813.

[96] Segre G, Campi R, Scarpellini F, Clavenna A, Zanetti M, Cartabia M, Bonati M. Interviewing children: the impact of the COVID-19 quarantine on children's perceived psychological distress and changes in routine. BMC Pediatr 2021;21:231. https://doi.org/10.1186/s12887-021-02704-1.

[97] Brooks SK, Webster RK, Smith LE, Woodland L, Wessely S, Greenberg N, Rubin GJ. The psychological impact of quarantine and how to reduce it: rapid review of the evidence. Lancet 2020;395:912–20. https://doi.org/10.1016/S0140-6736(20)30460-8.

[98] Uccella S, De Grandis E, De Carli F, D'Apruzzo M, Siri L, Preiti D, Di Profio S, Rebora S, Cimellaro P, Biolcati Rinaldi A, Venturino C, Petralia P, Ramenghi LA, Nobili L. Impact of the COVID-19 outbreak on the behavior of families in Italy: a focus on children and adolescents. Front Public Health 2021;9:32. https://doi.org/10.3389/fpubh.2021.608358.

Chapter 4

Complications: MISC and other complications

Lilia M. Sierra-Galan[1] **and Roberto M. Richheimer-Wohlmuth**[1,2]
[1]*Cardiology Department at the Cardiovascular Division of the American British Cowdray Medical Center, Mexico City, Mexico;* [2]*Paediatric Department of the American British Cowdray Medical Center, Mexico City, Mexico*

Introduction

Coronavirus disease 2019 (COVID-19) was initially identified in the pediatric age range (children and adolescents) in early January 2020, with much lower incidence and severity than adults [1]. In total, 80% of children will manifest any sign and/or symptom of COVID-19, which are usually mild and rarely fatal. The remaining 20% will not have clinical manifestations of the infection [1].

COVID-19 presentation can vary from asymptomatic, to the acute form, to the multisystem inflammatory syndrome in children (MIS-C), the long COVID, and death [2]. Interestingly, there is no homologous terminology for all different clinical presentations [3].

Acute-COVID-19

Definition

COVID-19 is an acute infection by the SARS-CoV-2 virus.

Different names

MIS-C definition. The earliest definition, by the Royal College of Pediatrics and Child Health (RCPCH), stated: a child presenting with persistent fever, inflammation (neutrophilia, elevated CRP, and lymphopenia), and evidence of single or multiorgan dysfunction (shock, cardiac, respiratory, renal, gastrointestinal, or neurological disorder) with additional features, which may include children fulfilling full or partial criteria for Kawasaki disease; exclusion of any other microbial cause and SARS-CoV-2 PCR testing may be positive or negative [4]. This was soon followed by the definitions by the Centers for Disease Control

Clinical Management of Pediatric COVID-19. https://doi.org/10.1016/B978-0-323-95059-6.00003-6

and Prevention (CDC): an individual aged <21 years presenting with fever, laboratory evidence of inflammation, and evidence of clinically severe illness requiring hospitalization, with multisystem ($\geq$2) organ involvement AND no alternative plausible diagnoses AND positive for current or recent SARS-CoV-2 infection by RT-PCR, serology, or antigen test; or COVID-19 exposure within the 4 weeks before the onset of symptoms [5], and the World Health Organization (WHO): children and adolescents 0–19 years old with fever >3 days AND two of the following: (a) rash or bilateral nonpurulent conjunctivitis or mucocutaneous inflammation signs; (b) hypotension or shock; (c) features of myocardial dysfunction, pericarditis, valvulitis, or coronary abnormalities; (d) evidence of coagulopathy; (e) acute gastrointestinal problems (diarrhea, vomiting, or abdominal pain) AND elevated markers of inflammation such as ESR, CRP, or procalcitonin AND no other obvious microbial cause of inflammation, including bacterial sepsis, staphylococcal or streptococcal shock syndromes AND evidence of COVID-19 (RT-PCR, antigen test or serology positive) or likely contact with patients with COVID-19 [6].

Clinical manifestations

Asymptomatic (19.3%), fever (59.1%), cough (55.9%), rhinorrhoea and nasal congestion (20.0%), myalgia and fatigue (18.7%), sore throat (18.2%) [3], shortness of breath and dyspnoea (11.7%), abdominal pain and diarrhea (6.5%), vomiting and nausea (5.4%), headache and dizziness (4.3%), pharyngeal erythema (3.3%), decreased oral intake (1.7%), rash (0.25%) [7–9], changes in smell [10], anosmia [3,7,11], and MIS-C.

Complications

Patchy lung damage (21.0% in chest X-ray (CXR) and 10.5% in CT), ground-glass opacity (6.0% in CXR and 32.9% in CT) or consolidation (2.4% in CXR and 6.5% in CT), requirement of mechanical ventilation (0.54%), shock (0.24%), disseminated intravascular coagulation (0.12%), kidney failure (0.12%), cardiac injury manifested as myocarditis, myopericarditis, pericarditis or pancarditis (0.10%), and MIS-C (0.14%) [7].

Long-COVID

Definition

According to the CDC, this includes a wide range of health consequences present four or more weeks after infection with SARS-CoV-2. However, the time frame could vary in the future when more evidence is gathered [12].

Different names

Post-COVID-19 conditions, long-COVID, long-haul COVID, postacute COVID, long-term effects of COVID-19, chronic COVID-19 [3].

Clinical manifestations

Anosmia (77.9%) [3,11], general symptoms such as fatigue (10.1%), insomnia (18.6%), nasal congestion/rhinorrhoea (12.4%), persistent muscle pain (10.1%), headache (10.1%), lack of concentration (10.1%), weight loss (7.7%), joint pain or swelling (6.9%), skin rashes (6.9%), chest tightness (6.2%), constipation (6.2%), persistent cough (5.4%), altered smell (4.6%), palpitations (3.8%), chest pain (3.1%), altered taste (3.1%), hypersomnia (3.1%), stomach/abdominal pain (2.3%), diarrhea (1.5%), menstruation disorders (1.5%) and others, such as dizziness, seizures, hallucinations, testicular pain (2.3%) [2,13,14], depression and anxiety symptoms [15,16], and late-onset MIS-C [17].

Incidence: >50% of children aged 6–16 years where these symptoms impair 42.6% during daily activities [2,3].

Complications

Dysphonia and dysphagia might be due to the postinflammatory condition, secondary to intubation, or both [3]. Dysfunctional/delayed bladder emptying, emotional lability, mild residual peripheral neuropathy, facial weakness, ptosis [18], acute disseminated encephalomyelitis, encephalitis, and Guillain–Barre Syndrome [19]. Palpitations (1.5%) and variations in heart rate (2.0%) [10], chest pain or tightness (3.1%–6.2%) [10], cardiac involvement, as ventricular dysfunction, coronary artery dilatation or aneurysms, arrhythmia, conduction abnormalities [20,21] and more rarely myocarditis, valvulitis, myopericarditis, pericarditis, or pancarditis [17,20,21]. Rhinorrhoea (52.4%) [11], persistent cough (1.0%–5.4%) [10,14], chest tightness, or difficult breathing [10]. Nausea, abdominal pain (5%) [11,22], diarrhea, and MIS-C [20,21].

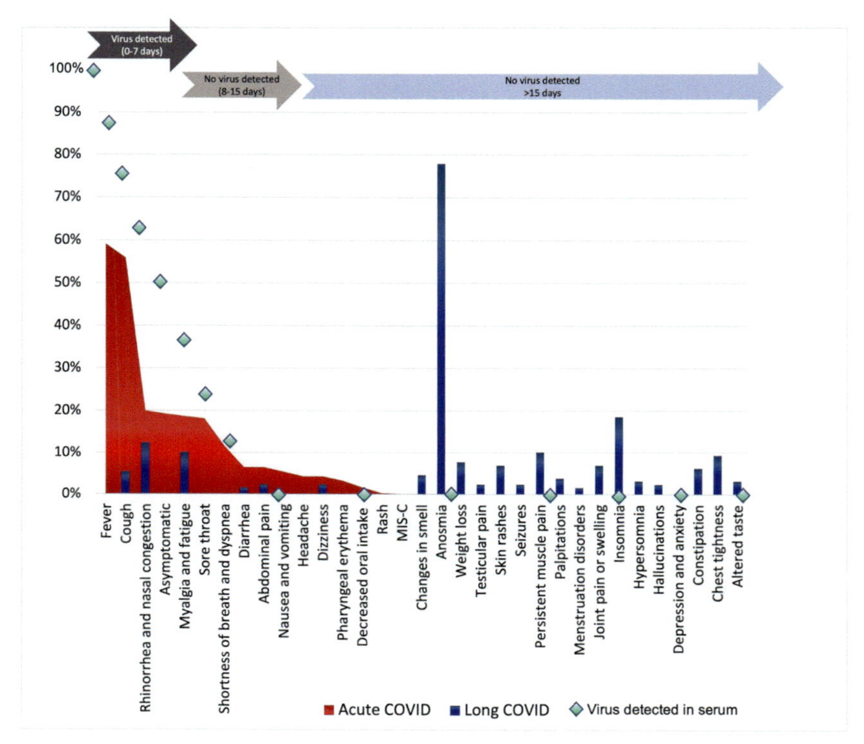

Graphical comparison of acute COVID and long COVID shows the prevalence of diverse clinical manifestations in acute and long COVID-19 in correlation with virus serum clearance. Most signs and symptoms are almost independent in acute and long-COVID states, with some minor overlapping [2,3,7−11,13−17]. The most symptomatic period coincides with the first 7 days on average (viremia period), with a progressive detection to almost complete clearance of the virus from plasma [23]. Based on this graphic representation, it can be theorized that the acute phase corresponds to a systemic reaction to the presence of the virus in the body. In contrast, the prolonged phase corresponds to a systemic reaction to an inflammation memory of the already cleared virus. Further research is needed to clarify these findings.

Pathophysiology of COVID-19 complications

SARS-CoV-2 virus damages type I and II pneumocytes. The ensuing inflammatory reaction breaks the epithelial barrier and the underlying vascular endothelium, allowing the migration of liquid and proteins into the alveolar space with a subsequent surfactant dysfunction, reduction of gas exchange surface, and an increase in obstructed areas, resulting in diminished gas exchange [24]. The cells of the olfactory neuroepithelium abundantly express the ACE-2 receptor, which acts as the SARS-CoV-2 receptor [25]. SARS-CoV-2 reaches the central nervous system through hematogenous dissemination. It stimulates inflammation, activates the coagulation cascade, and causes disarray of the production of *N*-methyl-D-aspartate and glutamate neurotransmitters, all of which result in neuronal damage [26]. In the heart, systemic inflammation produces a cytokine storm that directly injures myocytes and the vascular endothelium [17,27]. SARS-CoV2 also causes gastrointestinal, immune damage by migration of the CD4+ T-cells to the small intestine via the oral passage of the virus from the upper respiratory tract, which stimulates Th17 cells and neutrophils [28].

Multisystem inflammatory syndrome in children

MIS-C or pediatric inflammatory, multisystem syndrome (PIMS/PIMS-TS) is a rare systemic illness specific to the pediatric population, involving persistent fever and extreme inflammation, which appears to be an immune-mediated acute respiratory complication of SARS-CoV-2 infection [29−31].

Early in 2019, soon after the WHO declared COVID-19 a pandemic [32], it became clear that infection with the SARS-CoV-2 virus did not affect children as much as it did adults in the acute phase of the disease, except in children under one year, who are more prone to develop severe/critical illness due to bacterial pulmonary coinfection [33−37]. In March 2020, the spike protein (S) of the SARS-CoV-2 virus was shown to have an affinity for the angiotensin-converting enzyme 2 (ACE2) receptors [38]. Despite this lower severity of symptoms in children in the acute phase of COVID-19-19, late-onset severe cardiac disease related to COVID-19 in children was reported in May 2020 [39] and June 2020 [40,41]. This association, initially known as Paediatric multisystem inflammatory syndrome temporally associated with COVID-19(PIMS-Ts, or PIMS) [4] soon became known as MIS-C [29] and became known as one of the most severe forms of Sars-CoV-2-related disease in children and adolescents.

Epidemiology

MIS-C is an infrequent complication of COVID-19 in children [20], presenting in 1 in 3000−4000 children with COVID-19 [21,42], predominantly in those >5 years old, including teenagers, in contrast to <5 years in Kawasaki disease, with which MIS-C has extensive similarities. Proposed explanations for this difference include a better response in the earlier years to viral infections due to increased contact with viral infections and to MMR vaccines [43] and lower rates of SARS-CoV-2 infection in children due to the lesser expression of the cell surface enzyme ACE2, a receptor to which SARS-CoV-2 spike protein has been shown to attach to stimulate the entrance of the virus into human cells [36]. MIS-C affects males more than females (1.3−1.6: 1) [36,44]. Moreover, 79−95% of MIS-C occurs in previously healthy children, with asthma and obesity as the main comorbidities [20]. A study conducted in the United States found that Black, Hispanic or Latino, and Asian or Pacific Islander individuals are more prone to develop the syndrome [21].

Clinical manifestations

Presenting signs and symptoms with MIS-C varied by age and presence or absence of preceding COVID-19 [45]. While early into the pandemic some reports suggested that children infected with coronavirus disease SARS-CoV-2 present with a mild upper respiratory illness [8], and in a study of 171 children with confirmed COVID-19, only three required intensive care unit admission and only one death was observed [34], in early May 2020, investigators in the UK published a report describing eight severely ill pediatric patients presenting with hyperinflammatory shock with multiorgan involvement [39] high fever, rash, conjunctivitis, peripheral oedema, and gastrointestinal symptoms [46]. Detailed signs, symptoms, laboratory and imaging abnormalities are described above, under the RCPCH [4], the CDC [29] and the WHO [6] definitions of MIS-C. Teenagers and infants, especially those with preexisting cardiovascular disease, are at higher risk of morbidity and mortality, presenting with serious heart damage complications [47].

Figure (with permission) of a timeline of MIS-C presentation.

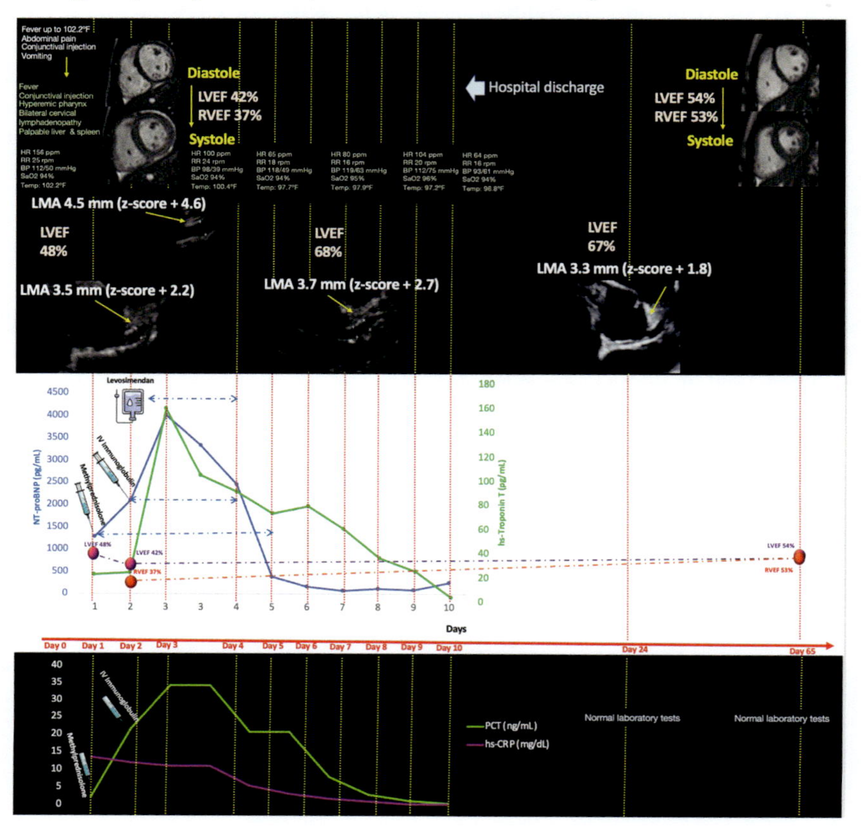

Abbreviations: *CMR*, cardiovascular magnetic resonance; *hs-CRP*, high-sensitive C-reactive protein; *hs-Troponin T*, high-sensitive Troponin T; *IV*, intravenous; *LMA*, left main coronary artery; *LVEF%*, left ventricular ejection fraction; *PCT*, procalcitonin; *RVEF%*, right ventricular ejection fraction. *Figure reproduced with written permission from Permanyer, publisher of Archivos de Cardiologia de Mexico from Ref.* [17].

This timeline is an example of the clinical course of a late presentation of the disease and its follow-up. The image represents the clinical signs and symptoms, vital signs, inflammation, cardiac dysfunction and damage, laboratory tests, echocardiographic measurements of LVEF% and left main coronary artery size, and CMR measurements of LVEF% and RVEF%, with the corresponding images of end-diastolic and end-systolic short-axis frames, including days when methylprednisolone, IV immunoglobulin, and levosimendan were administered. The red arrow represents the timeline of the patient's evolution from the development of the first clinical manifestations through day 65 when the control CMR was performed. Days are marked in red next to the arrow, from day 0 to day 10 during hospitalization until discharge, then days 24 and 65 of follow-up. Across the image (red on a white background and yellow on black background), the thin dashed lines mark and divide the days described in the timeline. According to the timeframe within the dashed lines, all clinical events, laboratory tests, and imaging findings are in the corresponding moments [17].

Causes, risk factors, and pathophysiology

A definitive explanation for the difference in the severity of symptoms in children compared to adults has proven elusive. Autoreactivity and post-SARS-CoV-2 exposure failure in regulation of the immune activation has been suggested as major causes of tissue damage [48]. Belay and coworkers identified the relationship between the geographic and temporal occurrence of peaks of COVID-19 followed 2–5 weeks later by surges in MIS-C cases [45]. Several authors hypothesized that MIS-C is due to delayed immunologic responses to infection by SARS-CoV-2 [2,6,21].

Some risk factors or associations have been described, such as the age range from 4 months to 17 years [49]; however, this association may change in the future when more evidence surfaces. In addition, males seem to be more often affected than females [46,50]. There are descriptions of higher prevalence in African, African-Caribbean, and Hispanic patients [46,50,51], which might be related to local identification and reporting instead of a true risk factor; more research is needed to clarify this data.

In July 2020, Lingappan et al. published various possible explanations for these differences. First, the pathophysiological mechanisms causing reduced pulmonary injury in children may comprise the decreased expression of the mediators, including ACE2 receptors, essential for viral access into the respiratory epithelium, which could prevent viral entry; next, the variances in the immune system responses in the young may include a relative dominance of CD4+ T cells, diminished neutrophil infiltration and production of proinflammatory cytokines, and augmented production of immunomodulatory cytokines in children compared with adults. Also, the developing lung in the young may have a greater capability to recuperate and repair after viral infection. The article also cites the possible explanation that the ratio of alveolar epithelial cells expressing ACE2 in adult lungs is higher compared with young lungs, which may cause a decreased viral access and copying in the pulmonary epithelial cells in children compared with adults, thus protecting the young partly by the diminished expression of receptors and other proteins that are vital for viral entry into the respiratory epithelium. There also seems to be a different equilibrium of pro- versus antiinflammatory cytokines in children and adults, with the proportion of antiinflammatory cytokines decreasing with age [36]. Abrehart et al. concluded that increasing age is the most reliable predictor of disease severity in COVID-19 but that the exact mechanisms supporting this observation are not entirely understood, leaving open the possibility that complex interactions between host factors and one or more SARS-CoV-2 binding factors may exist [52].

The eight patients in the communication by Riphagen et al. presented with hard to control fever up to 40°C, skin eruption, conjunctivitis, distal swelling, and widespread limb pain with substantial gastrointestinal discomfort and diarrhoea. All patients developed shock, were unresponsive to volume expanders, and required diverse vasopressors. However, most of the patients had no significant pulmonary problems and required mechanical ventilation support. Other findings included pleural, pericardial, and ascitic effusions indicative of widespread inflammatory activity. All patients tested negative for SARS-CoV-2 on respiratory tract aspirates. Lab markers indicative of inflammation (C-reactive protein, procalcitonin, ferritin, triglycerides, and D-dimers) were abnormally high, but no causative organism was found.

Initial EKGs were within normal limits. Echocardiograms frequently showed enhanced coronary artery echogenicity. Cardiac enzymes during the acute phase of the illness were elevated. The authors conclude that this array of signs, symptoms, and lab findings was a new phenomenon involving previously asymptomatic children with COVID-19, consisting of a hyper-inflammatory disorder with multiorgan participation similar to Kawasaki disease. Six of the eight children had either positive COVID-19 antibodies or had close interaction with identified COVID-19 patients [39]. Similar findings were reported by Verdoni et al., who noted in 10 patients a significant rise in incidence, age, cardiac participation, Kawasaki disease shock syndrome, and macrophage activation syndrome when comparing a group of pre-COVID-19 patients with a group of patients admitted during the initial phases of the pandemic [41].

Diagnosis

Diagnosis should include medical history, physical examination, and laboratory tests that meet the criteria stated in the definitions section (see above) [4–6]. Serial monitoring of laboratory markers of inflammation, an electrocardiogram, an echocardiogram, a baseline chest X-ray [4,5,53], and an abdominal ultrasound if severe abdominal pain is present [5] are recommended.

A Chest X-ray may miss some pulmonary infiltrates; therefore, a chest CT should be taken whenever possible [53].

EKG may show conduction disturbances, such as first, second, or third-degree A-V blocks, prolonged QT interval, repolarization abnormalities as

ST-segment elevation, and various ventricular arrhythmias [54,55]. An echocardiogram may reveal the general features of myocarditis and possibly coronary artery dilatation, valvulitis, and pericardial effusion similar to those in Kawasaki disease [17,41,54]; an abdominal ultrasound may reveal coexistent hepato-splenomegaly, bowel wall oedema, ascites, and lymphadenopathy [4,56].

Lastly, CMR has demonstrated in different case series and case reports the presence of diffuse myocardial oedema in tissue characterization and mapping techniques with or without the presence [17] of fibrosis or focal necrosis [57], consistent with postinfectious myocarditis [57].

Treatment

The objective of treatment is to reverse the shock state and organ dysfunction and prevent further injury and complications. Management by a multidisciplinary team is strongly recommended [58,59].

When a child meets MIS-C criteria (see earlier discussion), admission to the pediatric ICU is advised [60], because 73% [58] will require vasopressor and/or inotropic support.

Pharmacological treatment will vary on continuous individual assessment. Still, it will often include supportive care with oxygen administration, mechanical ventilation, fluid resuscitation, and vasopressor support according to established specific shock protocols, antiinflammatory and immunomodulatory therapies (intravenous immunoglobulin, methylprednisolone, anakinra, or levosimendan, antiplatelet and anticoagulation therapy) [4,17,49,59,61–65]. Extracorporeal membrane oxygenation might have to be considered if needed. The use of antibiotics must be tailored depending on specific needs, and remdesivir should be considered if SARS-CoV-2 remains positive [66].

Monitoring

Acute phase:

Laboratory tests (see specific tests in the definition section) [4,59].

Serial electrocardiograms at least every 48 h [59], at least one echocardiogram [54,59].

Chronic phase:

Follow-up echocardiogram 2 and 6 weeks, and one year after hospital discharge [59,62,67].

CMR at baseline, depending on the patient's respiratory and hemodynamic conditions; consider repeating 2–6 months after the acute phase, mainly in those patients with significant transient or persistent LV dysfunction [59].

Differential diagnosis

Disease	Feature in common	References
Kawasaki disease: • Partial Kawasaki • Kawasaki disease shock syndrome	Abdominal pain Thrombocytopenia Anemia Lymphopenia Elevated levels of ferritin, troponin, NT-proBNP, and D-dimer.	[41,51]
Scarlet fever	Rash Fever Lymphadenopathy	[68]
Toxic shock syndrome: • Nonstreptococcal • Streptococcal	Fever Rash (generalized erythroderma followed by desquamation for nonstreptococcal cases; cases with the streptococcal disease may desquamate in one or more sites) Hypotension Thrombocytopenia Central nervous system involvement Renal failure Multiorgan involvement ($\geq$3) Acute respiratory distress syndrome in streptococcal cases	[4,69,70]
Septic shock	Fever Rash (not typical) Septic shock hemodynamics	[71,72]
Rubella	Fever persisting for several days Conjunctival involvement Diffuse rash	[73]

Complications

Cardiovascular: mainly myocardial fibrosis, LV dysfunction, persistent pericarditis, and development of coronary aneurysms that pose a risk for thrombosis. This cardiovascular involvement usually resolves ad-integrum if patients are timely and optimally treated [17,74].

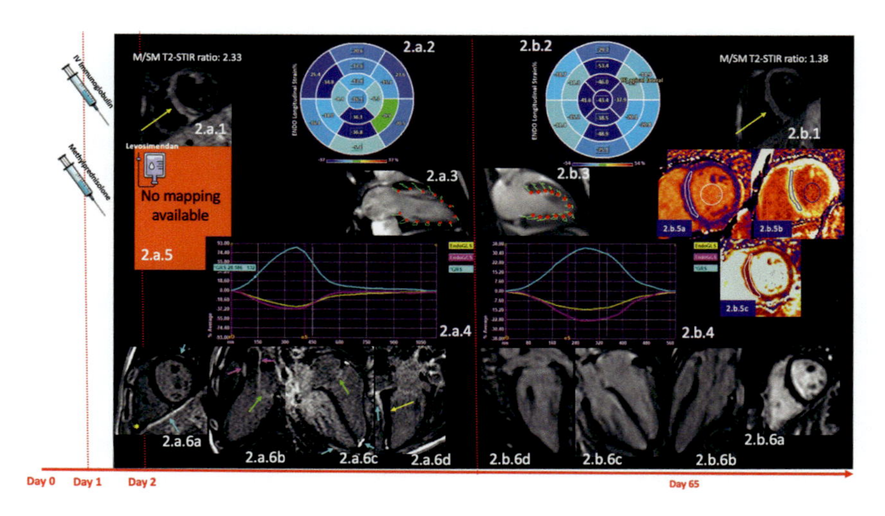

Abbreviations: *LV*, left ventricular; *M/SM*, myocardium/skeletal muscle, *T2-STIR*, T2-weighted with Short Tau Inversion Recovery. Reproduced with permission from Permanyer, publisher of Archivos de Cardiologia de Mexico from Ref. [17].

The figure compares the first CMR (**2.a**) study, performed on day one of hospitalization, and the follow-up CMR (**2.b**). **2.a.1** shows a T2-weighted STIR sequence in a short-axis view of the midventricle where the yellow arrow highlights the areas of increased signal intensity in the interventricular septum and a myocardium/skeletal muscle ratio of 2.33 corresponding to myocardial edema, that on its corresponding follow-up image (**2.b.1**) shows normal signal intensity with normal myocardium/skeletal muscle ratio, suggesting the absence of myocardial edema. Image **2.a.2, 3**, and **4** corresponds to the functional and strain analysis showing regional wall motion abnormalities and the reduced LV global function with a global strain of 29.20%, which on its corresponding follow-up images (**2.b.2,3** and **4**) shows a normal LV global function with a global strain of 46.49%. Images **2.a.5.** are the mapping techniques unavailable for the first CMR study. The corresponding images **2.b.5a, b**, and **c** show the mapping analysis, figure **2.b.5a** shows the native T1 of 1071 ms (normal for the 1.5 T magnet used), figure **2.b.5b** shows the postgadolinium map, the extracellular volume obtained was 26% which is normal, figure **2.b.5c** shows the T2 mapping with a native T2 value of 48 ms (normal for the 1.5 T magnet used). At the bottom of the figure, images **2a.6** and **2b.6** show the late gadolinium-enhanced T1-weighted inversion-recovery with phase-sensitive reconstruction algorithm sequences, **2.a.6a** is the short-axis view at midventricle, **2.a.6b**, is the 3-chambers view, **2.a.6c** is the 4-chambers view, and the **2.a.6d** is the two chambers view of the first CMR to the left side of the figure, that nicely show linear intramural and patchy enhancement areas in different locations in the myocardium as marked by the yellow arrow in image **2.a.6d**; as well as enhancement of the valves (green arrows) and ascending aorta in image **2a.6b** (pink arrows), and the enhancement of the pericardium (blue arrows) with mild pericardial effusion (yellow asterisk). On the right side of the figure, the corresponding images and locations of the second study on day 65 of follow-up (images **2.b.6a, 2b.6b, 2.b.6c**, and **2.b.6d**) show no myocardial, valvular, aortic, or pericardial enhancement and no pericardial effusion.

Unfortunately, not all centers have the ability to provide the necessary support, which can lead to serious morbidity and death reported [49].

The long-term prognosis of survivors is yet unknown.

Conclusion

MIS-C is a rare but potentially fatal, severe complication of COVID-19 in children and adolescents. Therefore, early recognition of its features is essential for timely and adequate therapy, which requires intensive treatment by a highly trained multidisciplinary team to ensure good results.

References

[1] Borrelli M, Corcione A, Castellano F, Fiori Nastro F, Santamaria F. Coronavirus disease 2019 in children. Front Pediatr 2021;9:668484. https://doi.org/10.3389/fped.2021.668484.

[2] Thomson H. Children with long covid. New Sci 2021;249(3323):10–1. https://doi.org/10.1016/s0262-4079(21)00303-1.

[3] Thallapureddy K, Thallapureddy K, Zerda E, et al. Long-term complications of COVID-19 infection in adolescents and children. Curr Pediatr Rep 2022;10(1):11–7. https://doi.org/10.1007/s40124-021-00260-x.

[4] RCPCH. Guidance: paediatric multisystem inflammatory syndrome temporally associated with COVID-19.

[5] CDC. Centers for Disease Control and Prevention. Emergency preparedness and response: health alert network. https://emergency.cdc.gov/han/index.%0Dasp. Accessed June 22, 2021.

[6] Freedman, S.; Godfred-Cato, S.; Gorman, R.; Lodha, R.; Mofenson, L., Murthy, S.; Semple, C.; Sigfrid, L.; Whittaker E. Multisystem inflammatory syndrome in children and adolescents with COVID-19 Scientific brief.

[7] Hoang A, Chorath K, Moreira A, et al. COVID-19 in 7780 pediatric patients: a systematic review. EClinicalMedicine 2020;24:100433. https://doi.org/10.1016/j.eclinm.2020.100433.

[8] Castagnoli R, Votto M, Licari A, et al. Severe acute respiratory syndrome coronavirus 2 (SARS-CoV-2) infection in children and adolescents: a systematic review. JAMA Pediatr 2020;174(9):882–9. https://doi.org/10.1001/jamapediatrics.2020.1467.

[9] Galindo R, Chow H, Rongkavilit C. COVID-19 in children: clinical manifestations and pharmacologic interventions including vaccine trials. Pediatr Clin North Am 2021;68(5):961–76. https://doi.org/10.1016/j.pcl.2021.05.004.

[10] Osmanov IM, Spiridonova E, Bobkova P, et al. Risk factors for long covid in previously hospitalised children using the ISARIC Global follow-up protocol: a prospective cohort study. Eur Respir J 2021. https://doi.org/10.1183/13993003.01341-2021.

[11] Molteni E, Sudre CH, Canas LS, et al. Illness duration and symptom profile in symptomatic UK school-aged children tested for SARS-CoV-2. Lancet Child Adolesc Health 2021;5(10):708–18. https://doi.org/10.1016/S2352-4642(21)00198-X.

[12] CDC. Post-COVID conditions: information for healthcare providers. July 19, 2021. Published 2021, https://www.cdc.gov/coronavirus/2019-ncov/hcp/clinical-care/post-covid-conditions.html?CDC_AA_refVal=https%3A%2F%2Fwww.cdc.gov%2Fcoronavirus%2F2019-ncov%2Fhcp%2Fclinical-care%2Flate-sequelae.html.

[13] Lopez-Leon S, Wegman-Ostrosky T, Perelman C, et al. More than 50 long-term effects of COVID-19: a systematic review and meta-analysis. Sci Rep 2021;11(1):16144. https://doi.org/10.1038/s41598-021-95565-8.

[14] Buonsenso D, Munblit D, De Rose C, et al. Preliminary evidence on long COVID in children. Acta Paediatr 2021;110(7):2208–11. https://doi.org/10.1111/apa.15870.

[15] Racine N, McArthur BA, Cooke JE, Eirich R, Zhu J, Madigan S. Global prevalence of depressive and anxiety symptoms in children and adolescents during COVID-19: a meta-analysis. JAMA Pediatr 2021;175(11):1142–50. https://doi.org/10.1001/jamapediatrics.2021.2482.

[16] Danese A, Smith P, Chitsabesan P, Dubicka B. Child and adolescent mental health amidst emergencies and disasters. Br J Psychiatry 2020;216(3):159–62. https://doi.org/10.1192/bjp.2019.244.

[17] Richheimer-Wohlmuth RM, González-López A, Salcido-Carmona CA, Sierra-Galán LM. Late presentation of multisystem inflammatory syndrome in children associated with SARS-CoV-2 infection, Kawasaki disease, pancarditis, and great vessels arteritis. Arch Cardiol Mex February 2022. https://doi.org/10.24875/ACM.21000264.

[18] Lindan CE, Mankad K, Ram D, et al. Neuroimaging manifestations in children with SARS-CoV-2 infection: a multinational, multicentre collaborative study. Lancet Child Adolesc Health 2021;5(3):167–77. https://doi.org/10.1016/S2352-4642(20)30362-X.

[19] Singer TG, Evankovich KD, Fisher K, Demmler-Harrison GJ, Risen SR. Coronavirus infections in the nervous system of children: a scoping review making the case for long-term neurodevelopmental surveillance. Pediatr Neurol 2021;117:47–63. https://doi.org/10.1016/j.pediatrneurol.2021.01.007.

[20] Alsaied T, Tremoulet AH, Burns JC, et al. Review of cardiac involvement in multisystem inflammatory syndrome in children. Circulation 2021;143(1):78–88. https://doi.org/10.1161/CIRCULATIONAHA.120.049836.

[21] Payne AB, Gilani Z, Godfred-Cato S, et al. Incidence of multisystem inflammatory syndrome in children among US persons infected with SARS-CoV-2. JAMA Netw Open 2021;4(6). https://doi.org/10.1001/jamanetworkopen.2021.16420. e2116420-e2116420.

[22] Radtke T, Ulyte A, Puhan MA, Kriemler S. Long-term symptoms after SARS-CoV-2 infection in children and adolescents. JAMA 2021;326(9):869–71. https://doi.org/10.1001/jama.2021.11880.

[23] Hagman K, Hedenstierna M, Rudling J, et al. Duration of SARS-CoV-2 viremia and its correlation to mortality and inflammatory parameters in patients hospitalized for COVID-19: a cohort study. Diagn Microbiol Infect Dis 2022;102(3):115595. https://doi.org/10.1016/j.diagmicrobio.2021.115595.

[24] Oliveira TL, Melo IS, Cardoso-Sousa L, et al. Pathophysiology of SARS-CoV-2 in lung of diabetic patients. Front Physiol 2020;11:587013. https://doi.org/10.3389/fphys.2020.587013.

[25] Butowt R, von Bartheld CS. Anosmia in COVID-19: underlying mechanisms and assessment of an olfactory route to brain infection. Neurosci Rev J Bringing Neurobiol Neurol Psychiatry 2021;27(6):582–603. https://doi.org/10.1177/1073858420956905.
[26] Boldrini M, Canoll PD, Klein RS. How COVID-19 affects the brain. JAMA Psychiatr 2021;78(6):682–3. https://doi.org/10.1001/jamapsychiatry.2021.0500.
[27] Basu-Ray I, Almaddah NK, Adeboye AA. Cardiac manifestations of coronavirus (COVID-19). In: StatPearls [Internet]. Treasure Island (FL): StatPearls Publishing; January 2022. https://www.ncbi.nlm.nih.gov/books/NBK556152/. [Accessed 3 April 2022].
[28] Ye Q, Wang B, Zhang T, Xu J, Shang S. The mechanism and treatment of gastrointestinal symptoms in patients with COVID-19. Am J Physiol Gastrointest Liver Physiol 2020;319(2):G245–52. https://doi.org/10.1152/ajpgi.00148.2020.
[29] CDC. Centers for Disease Control. Multisystem inflammatory syndrome in children (MIS-C) associated with coronavirus disease 2019 (COVID-19). https://emergency.cdc.gov/han/2020/han00432.asp. [Accessed 3 April 2022].
[30] Harahsheh AS, Sharron MP, Bost JE, Ansusinha E, Wessel D, DeBiasi RL. Comparison of first and second wave cohorts of multisystem inflammatory disease syndrome in children. Pediatr Infect Dis J 2022;41(1):e21–5. https://doi.org/10.1097/INF.0000000000003388.
[31] Rhys-Evans S. Call for a universal PIMS-TS/MIS-C case definition. Arch Dis Child 2022;107(3):e10. https://doi.org/10.1136/archdischild-2021-322829.
[32] Ghebreyesus TA. WHO Director-General's opening remarks at the media briefing on COVID-19 - 11 March 2020. March 11, 2020. https://www.who.int/director-general/speeches/detail/who-director-general-s-opening-remarks-at-the-media-briefing-on-covid-19—11-march-2020. Published 2020. Accessed April 3, 2022.
[33] Dong Y, Mo X, Hu Y, et al. Epidemiology of COVID-19 among children in China. Pediatrics 2020;145(6). https://doi.org/10.1542/peds.2020-0702.
[34] Lu X, Zhang L, Du H, et al. SARS-CoV-2 infection in children. N Engl J Med 2020;382(17):1663–5. https://doi.org/10.1056/NEJMc2005073.
[35] Shekerdemian LS, Mahmood NR, Wolfe KK, et al. Characteristics and outcomes of children with coronavirus disease 2019 (COVID-19) infection admitted to US and Canadian pediatric intensive care units. JAMA Pediatr 2020;174(9):868–73. https://doi.org/10.1001/jamapediatrics.2020.1948.
[36] Lingappan K, Karmouty-Quintana H, Davies J, Akkanti B, Harting MT. Understanding the age divide in COVID-19: why are children overwhelmingly spared? Am J Physiol Lung Cell Mol Physiol 2020;319(1):L39–44. https://doi.org/10.1152/ajplung.00183.2020.
[37] Xia W, Shao J, Guo Y, Peng X, Li Z, Hu D. Clinical and CT features in pediatric patients with COVID-19 infection: different points from adults. Pediatr Pulmonol 2020;55(5):1169–74. https://doi.org/10.1002/ppul.24718.
[38] Wrapp D, Wang N, Corbett KS, et al. Cryo-EM structure of the 2019-nCoV spike in the prefusion conformation. Science 2020;367(6483):1260–3. https://doi.org/10.1126/science.abb2507.
[39] Riphagen S, Gomez X, Gonzalez-Martinez C, Wilkinson N, Theocharis P. Hyperinflammatory shock in children during COVID-19 pandemic. Lancet (London, England) 2020;395(10237):1607–8. https://doi.org/10.1016/S0140-6736(20)31094-1.
[40] Jones VG, Mills M, Suarez D, et al. COVID-19 and Kawasaki disease: novel virus and novel case. Hosp Pediatr 2020;10(6):537–40. https://doi.org/10.1542/hpeds.2020-0123.
[41] Verdoni L, Mazza A, Gervasoni A, et al. An outbreak of severe Kawasaki-like disease at the Italian epicentre of the SARS-CoV-2 epidemic: an observational cohort study. Lancet

(London, England) 2020;395(10239):1771–8. https://doi.org/10.1016/S0140-6736(20)31103-X.

[42] Holm M, Hartling UB, Schmidt LS, et al. Multisystem inflammatory syndrome in children occurred in one of four thousand children with severe acute respiratory syndrome coronavirus 2. Acta Paediatr 2021;110(9):2581–3. https://doi.org/10.1111/apa.15985.

[43] Matucci-Cerinic C, Caorsi R, Consolaro A, Rosina S, Civino A, Ravelli A. Multisystem inflammatory syndrome in children: unique disease or part of the Kawasaki disease spectrum? Front Pediatr 2021;9. https://doi.org/10.3389/fped.2021.680813.

[44] Abrams JY, Godfred-Cato SE, Oster ME, et al. Multisystem inflammatory syndrome in children associated with severe acute respiratory syndrome coronavirus 2: a systematic review. J Pediatr 2020;226:45–54.e1. https://doi.org/10.1016/j.jpeds.2020.08.003.

[45] Belay ED, Abrams J, Oster ME, et al. Trends in geographic and temporal distribution of US children with multisystem inflammatory syndrome during the COVID-19 pandemic. JAMA Pediatr 2021;175(8):837–45. https://doi.org/10.1001/jamapediatrics.2021.0630.

[46] Ahmed M, Advani S, Moreira A, et al. Multisystem inflammatory syndrome in children: a systematic review. EClinicalMedicine 2020;26:100527. https://doi.org/10.1016/j.eclinm.2020.100527.

[47] Cuppari C, Ceravolo G, Ceravolo MD, et al. Covid-19 and cardiac involvement in childhood: state of the art. J Biol Regul Homeost Agents 2020;34(4 Suppl. 2):121–5 [Special issue: focus on pediatric cardiol].

[48] Jiang L, Tang K, Levin M, et al. COVID-19 and multisystem inflammatory syndrome in children and adolescents. Lancet Infect Dis 2020;20(11):e276–88. https://doi.org/10.1016/S1473-3099(20)30651-4.

[49] European Centre for Disease Prevention and Control. Paediatric inflammatory multisystem syndrome and SARS-CoV-2 infection in children. Stockholm.

[50] Godfred-Cato S, Bryant B, Leung J, et al. COVID-19-Associated multisystem inflammatory syndrome in children - United States, March-July 2020. MMWR Morb Mortal Wkly Rep 2020;69(32):1074–80. https://doi.org/10.15585/mmwr.mm6932e2.

[51] Vincent, S.BM, Ermias B, Michael L, James S, Marconi. CDC center for preparedness and response: multisystem inflammatory syndrome in children (MIS-C) associated with coronavirus disease 2019 (COVID- 19). In: Clinician outreach and communication activity (COCA) webinar. CDC website.

[52] Abrehart T, Suryadinata R, McCafferty C, et al. Age-related differences in SARS-CoV-2 binding factors: an explanation for reduced susceptibility to severe COVID-19 among children? Paediatr Respir Rev 2022. https://doi.org/10.1016/j.prrv.2022.01.008.

[53] Hameed S, Elbaaly H, Reid CEL, et al. Spectrum of imaging findings at chest radiography, US, CT, and MRI in multisystem inflammatory syndrome in children associated with COVID-19. Radiology 2021;298(1):E1–10. https://doi.org/10.1148/radiol.2020202543.

[54] Belhadjer Z, Méot M, Bajolle F, et al. Acute heart failure in multisystem inflammatory syndrome in children in the context of global SARS-CoV-2 pandemic. Circulation 2020;142(5):429–36. https://doi.org/10.1161/CIRCULATIONAHA.120.048360.

[55] Toubiana J, Poirault C, Corsia A, et al. Kawasaki-like multisystem inflammatory syndrome in children during the covid-19 pandemic in Paris, France: prospective observational study. BMJ 2020;369:m2094. https://doi.org/10.1136/bmj.m2094.

[56] Clinician outreach and communication activity (COCA) webinar. CDC center for preparedness and response: multisystem inflammatory syndrome in children (MIS-C) associated with coronavirus disease 2019 (COVID- 19). CDC website. Chrome-extension://efaidnbmnnnibpcajpcglclefindmkaj/viewer.html?pdfurl=https%3A%2F%2Femergency.cd

c.gov%2Fcoca%2Fppt%2F2020%2FCOCA_Call_Slides_05_19_2020.pdf&clen=3545831&chunk=true.
[57] Blondiaux E, Parisot P, Redheuil A, et al. Cardiac MRI in children with multisystem inflammatory syndrome associated with COVID-19. Radiology 2020;297(3):E283−8. https://doi.org/10.1148/radiol.2020202288.
[58] Belot A, Antona D, Renolleau S, et al. SARS-CoV-2-related paediatric inflammatory multisystem syndrome, an epidemiological study, France, 1 March to 17 May 2020. Euro Surveill Bull Eur sur les Mal Transm = Eur Commun Dis Bull. 2020;25(22). https://doi.org/10.2807/1560-7917.ES.2020.25.22.2001010.
[59] Henderson LA, Canna SW, Friedman KG, et al. American College of rheumatology clinical guidance for multisystem inflammatory syndrome in children associated with SARS-CoV-2 and hyperinflammation in pediatric COVID-19: version 2. Arthritis Rheumatol 2021;73(4):e13−29. https://doi.org/10.1002/art.41616.
[60] Statement to the Media following the 2 May Pediatric Intensive Care-COVID-19 International Collaborative Conference Call. Kawasaki Disease Foundation. https://kdfoundation.org/statement-to-the-media-following-the-2-may-pediatric-intensive-care-covid-19-international-collaborative-conference-call/.
[61] Multisystem Inflammatory Syndrome (MIS). Information for Healthcare Providers about Multisystem Inflammatory Syndrome in Children (MIS-C). CDC website. https://www.cdc.gov/mis/mis-c/hcp/index.html.
[62] Whittaker E, Bamford A, Kenny J, et al. Clinical characteristics of 58 children with a pediatric inflammatory multisystem syndrome temporally associated with SARS-CoV-2. JAMA 2020;324(3):259−69. https://doi.org/10.1001/jama.2020.10369.
[63] American Academy of Pediatrics. What is the case definition of multisystem inflammatory syndrome in children (MIS-C)? https://www.aap.org/en/pages/2019-novel-coronavirus-covid-19-infections/clinical-guidance/multisystem-inflammatory-syndrome-in-children-mis-c-interim-guidance/. Accessed April 9, 2022.
[64] Brissaud O, Botte A, Cambonie G, et al. Experts' recommendations for the management of cardiogenic shock in children. Ann Intensive Care 2016;6(1):14. https://doi.org/10.1186/s13613-016-0111-2.
[65] Weiss SL, Peters MJ, Alhazzani W, et al. Surviving sepsis campaign international guidelines for the management of septic shock and sepsis-associated organ dysfunction in children. Pediatr Crit Care Med J Soc Crit Care Med World Fed Pediatr Intensive Crit Care Soc 2020;21(2):e52−106. https://doi.org/10.1097/PCC.0000000000002198.
[66] Harwood R, Allin B, Jones CE, et al. A national consensus management pathway for paediatric inflammatory multisystem syndrome temporally associated with COVID-19 (PIMS-TS): results of a national Delphi process. Lancet Child Adolesc Health 2021;5(2):133−41. https://doi.org/10.1016/S2352-4642(20)30304-7.
[67] Chiotos K, Bassiri H, Behrens EM, et al. Multisystem inflammatory syndrome in children during the coronavirus 2019 pandemic: a case series. J Pediatr Infect Dis Soc 2020;9(3):393−8. https://doi.org/10.1093/jpids/piaa069.
[68] Basetti S, Hodgson J, Rawson TM, Majeed A. Scarlet fever: a guide for general practitioners. Lond J Prim Care 2017;9(5):77−9. https://doi.org/10.1080/17571472.2017.1365677.
[69] National Notifiable Diseases Surveillance System (NNDSS). Toxic Shock Syndrome (Other Than Streptococcal) (TSS) 1997 Case Definition. CDC website. https://ndc.services.cdc.gov/case-definitions/toxic-shock-syndrome-1997/. Accessed April 9, 2022.

[70] National Notifiable Diseases Surveillance System (NNDSS). Streptococcal toxic shock syndrome (STSS) (Streptococcus pyogenes) 2010 case definition. CDC website.

[71] Kawasaki T. Update on pediatric sepsis: a review. J Intens care 2017;5:47. https://doi.org/10.1186/s40560-017-0240-1.

[72] Garcia PCR, Tonial CT, Piva JP. Septic shock in pediatrics: the state-of-the-art. J Pediatr (Rio J). 2020;96(Suppl. 1):87–98. https://doi.org/10.1016/j.jped.2019.10.007.

[73] Winter AK, Moss WJ. Rubella. Lancet (London, England) 2022;399(10332):1336–46. https://doi.org/10.1016/S0140-6736(21)02691-X.

[74] Webster G, Patel AB, Carr MR, et al. Cardiovascular magnetic resonance imaging in children after recovery from symptomatic COVID-19 or MIS-C: a prospective study. J Cardiovasc Magn Reson 2021;23(1):86. https://doi.org/10.1186/s12968-021-00786-5.

Chapter 5

Diagnosis of pediatric COVID-19

Joseph L. Mathew and Ketan Kumar
Department of Pediatrics, Advanced Pediatrics Centre, Postgraduate Institute of Medical Education and Research (PGIMER), Chandigarh, Punjab, India

Introduction

Since the detection of the first case in Wuhan, China in December 2019 [1], the COVID-19 pandemic has proven to be one of the most challenging problems faced in modern times. Despite more than 2 years since its onset, during which the entire globe has suffered, and millions of lives have been lost, the end is still not within sight. A definite cure for the disease remains elusive, hence treatment is predominantly supportive, and prevention appears to be the best hope.

Against this background, accurate diagnosis of COVID-19 assumes great importance because it has implications not only for the management of individual patients but also for the implementation of appropriate infection control measures. The latter includes initiation and termination of isolation for index cases, contact tracing, and quarantine for contacts. Rapid and accurate detection of infectivity and reliable confirmation of noninfectivity is of paramount importance for judicious utilization of scarce and overstretched healthcare resources. In infected people, markers that can predict clinical deterioration, need for escalation of therapy, or development of complications are of great importance. At the same time, formulation of policies for population-based vaccination, restriction of movement, etc. needs evaluation of population susceptibility, or in other words, evidence of immunity against the pathogen.

Laboratory tests to diagnose COVID-19 in human beings can be broadly categorized as direct and indirect tests. Direct tests are those which demonstrate the actual presence of the virus (or its genetic or antigenic material) in biological specimens. Examples include reverse transcriptase polymerase chain reaction (RT-PCR) and immunoassays to detect viral antigens. On the other hand, indirect tests merely reflect changes in human tissues or organs in

Clinical Management of Pediatric COVID-19. https://doi.org/10.1016/B978-0-323-95059-6.00005-X

response to the virus, but these could also be found in other infective and noninfective disease states. This group of investigations includes assays for antibodies against viral proteins, various inflammatory markers, for example, leukocyte counts, C-reactive protein (CRP), etc. They also include radiological investigations including chest X-rays and computed tomography (CT) scans [2,3].

Before discussing COVID-19 diagnosis in children, it is important to consider why this could be different from that in adults. It is now well established that the virus tends to cause milder symptoms and less severe disease in the pediatric population due to biological and epidemiological differences. The former includes variation in immune system function and responses, and less frequent presence of comorbidities associated with higher risk [4]. The latter includes less mobility and lower risk of exposure. The concern that asymptomatic (or less symptomatic) infection in children could make them silent spreaders of the infection has not been validated [5]. Initially, there was speculation that the viral load in infected children may be lower than among adults. However, studies have failed to show any difference in the viral load present in the upper respiratory tract of children as compared with adults, in both asymptomatic and symptomatic subjects [6,7]. For these reasons, the direct methods to diagnose COVID-19 in adults can be used in children also. Although obtaining a satisfactory respiratory specimen is challenging in infants and young children, there are currently no guidelines that suggest a different approach in them.

Last but not the least, the virus itself, and our knowledge about it, continue to evolve at a rapid pace. Therefore, the practices being followed currently are "best practices" in the light of available scientific evidence, but might easily become redundant in the future.

Direct tests

The direct tests are based on the detection and demonstration of intact viable viral particles or their components in biological specimens of children. Currently, these tests are the mainstay of diagnosis. Under most circumstances, a positive result would mean SARS-CoV-2 infection. Thus, these tests have a high positive predictive value. However, the sensitivity varies based on the methods employed for specimen collection, processing, and laboratory procedures. Therefore, stringent specimen collection, optimum storage and processing conditions, and appropriate lab procedures with special attention to biosafety precautions are necessary to correctly interpret the test results.

Nucleic acid amplification test

Currently, nucleic acid amplification tests (NAATs) are considered to be the most reliable among the diagnostic tests available for clinical use. These are based on the amplification of the viral ribonucleic acid (RNA) and its detection in biological specimens of children. As a consequence of the amplification step, they have higher sensitivity than other tests. While RT-PCR is the most common NAAT being used worldwide, there are other methods in the same category. With time, point-of-care (POC) NAATs, some of which can even be self-administered, have become available and have received the United States Food and Drug Administration approval. These are minimally invasive tests in the sense that most involve obtaining a swab from the nasopharynx or other upper respiratory sites. However, lab-based NAATs continue to be considered more robust [8,9].

The RT-PCR process for identifying viral pathogens involves a common set of steps. Viral RNA is extracted from biological specimens, transcribed into complementary deoxy-ribonucleic acid, followed by amplification of specific parts, and conversion into quantifiable results using primers and probes. In short, extracted RNA is added into a mixture containing enzymes and reagents, in a thermocycler, under strict reaction conditions. Although different RT-PCR assays may differ in the gene targets chosen for amplification, the reaction conditions are standardized by the US Centers for Disease Control and Prevention (CDC) [3]. In general, it is recommended to have two independent targets on the viral genome to optimize the test performance. Commonly employed targets include regions on the SARS-CoV2 viral envelope (E), nucleocapsid (N), spike (S), and RNA-dependent RNA polymerase genes [8].

Although RT-PCR is currently considered the gold standard for COVID-19 diagnosis, it does not always have the desired accuracy. Diagnostic accuracy studies are hampered by the absence of a definitive test for comparison, as well as heterogeneity in reported studies [10]. Using a latent-class analysis model, one study reported RT-PCR sensitivity of only 68%, although specificity was 99% [11]. A systematic review including 20 diagnostic test studies reported overall sensitivity of 96%, meaning that 4% of cases would be missed [12]. To minimize the possibility of missing cases, repeating the test after 24 h has been suggested [13].

As it is unlikely that RT-PCR will be replaced by a more accurate test in the near future, it is imperative that decisions regarding initiation and termination of isolation of patients are made, taking into consideration their pretest probability (of having or not having the infection) and the consequences of the decision being made.

Diagnostic performance of RT-PCR is also influenced by the characteristics of the biological specimen being tested. In children, a combined nasopharyngeal (NP) and oropharyngeal (OP) swab is preferred. Alternatively, separately taken swabs may be put into the same container. If neither is feasible, a single NP swab should be taken. Collection should be done using dacron or polyester flocked swabs in a viral transport medium (VTM), which can be stored at 2–8°C for up to 12 days. However, all efforts should be made to transport the specimen to the testing lab as soon as possible [8].

If testing is required later in the disease course, or if the upper respiratory tract specimen tests negative despite strong clinical suspicion, a specimen from the lower respiratory tract (LRT) may be considered. This includes spontaneously expectorated or induced sputum, endotracheal aspirate, or bronchoalveolar lavage fluid. Although these specimens might offer higher yields, it is generally difficult to obtain them in children. Further, there is a higher risk of aerosolization with such procedures. Hence, LRT specimens are obtained only in a limited number of situations [14]. To the best of our knowledge, there are no studies in children, comparing the diagnostic yield in simultaneously collected upper versus lower respiratory tract specimens.

Some studies suggested that saliva obtained during cough, or even collected by passive drooling, could have acceptable accuracy compared to NP/OP swabs [15–17]. Saliva seems an attractive specimen in the pediatric age group, considering that its collection is noninvasive and can be done passively. However, currently, the World Health Organization (WHO) does not support the use of saliva alone as a diagnostic specimen [8]. Fecal specimens have been reported to be positive for viral RNA, for a longer duration than respiratory tract samples; hence can be tried [18,19]. The use of both saliva and fecal specimens may reduce the need for protection against aerosolization. The SARS-CoV-2 virus genetic material has also been detected in the blood, urine, semen, and cerebrospinal fluid of patients; however, these specimens are not currently recommended for diagnostic purposes.

An important consideration while testing for COVID-19 is the timing of the test. RT-PCR is expected to be positive in a patient harboring the virus, approximately 1–3 days before the onset of symptoms. The maximum probability of detection is around the time of symptom onset [8]. Thereafter, the rate of detection from upper respiratory specimens continues to decline. It may be prolonged in patients with more severe illnesses compared to those with milder symptoms. Viral nucleic acid is detectable for a longer duration in lower respiratory and fecal specimens, compared to upper respiratory specimens.

> Early evidence suggests that the Omicron variant of SARS-CoV-2 preferentially infects the upper respiratory tract cells compared to those of the lower respiratory tract. Whether or not it translates into a better diagnostic yield from upper respiratory tract samples is debatable.

It is important to remember that the mere detection of viral genetic material in biological specimens does not necessarily indicate the presence of a viable virus or infectivity of the patient. Early during the pandemic, the criteria for patients to be declared "COVID negative" were based on two negative RT-PCR tests, at an interval of 24 h. For this reason, several asymptomatic children remained in hospital (or isolation) for several days/weeks awaiting two negative test reports. Owing to the recognition that test positivity does not always correlate with infectivity, there is a shift from a test-based strategy to a symptom-based strategy for decisions on ending patient isolation [9,20,21].

In addition to qualitative RT-PCR (i.e., reporting test results as positive or negative), some tests also report the cycle threshold (Ct) value. By definition, Ct value is the number of amplification cycles required for the fluorescence of the PCR product to cross the detection threshold. Thus, the Ct value is inversely related to the amount of viral genetic material. In general, a change in Ct value by 1 represents a twofold change in the viral load. However, the Ct values cannot be taken as absolute correlates of infectivity because infectivity requires intact viral particles. There is evidence to suggest that a lower Ct value is associated with poorer patient outcomes [22,23]. A Ct value cut-off of ≥36 has been used to label a sample as nondiagnostic [24]. However, since the values are not comparable between assays and any significance can only be assigned to them with reference to a particular lab, there is no absolute universal cut-off for clinical use [25].

Apart from RT-PCR there are other NAATs, some of which have been approved. These are based on techniques such as loop-mediated isothermal amplification (LAMP), helicase-dependent amplification, clustered regularly interspaced short palindromic repeats, and others. As some of these techniques do not require as stringent conditions as lab-based NAATs, they can be used in commercially available kits, sometimes at the bedside. These kits are usually validated against a lab-based RT-PCR, carry specific instructions for use, and may need verification with a more reliable method in case of doubtful results [3]. One specific example is that of the cartridge-based Xpert Xpress SARS-CoV-2 test, which works on a fully automated system, thus reducing the risk of contamination and exposure to lab personnel. The turnaround time is 45 min. It uses the same platform approved by the WHO for the diagnosis of tuberculosis (GeneXpert MTB/RIF) [26].

A number of functional variants of SARS-CoV-2 have emerged during the pandemic, some with enhanced transmission potential and ability to evade immune responses. While detection of SARS-CoV-2 genetic material is sufficient to make a diagnosis of COVID-19 infection, identification of the specific variant requires further characterization of the genetic material. Next-generation sequencing is the gold standard for variant identification. However, it is a cost, time, and labor-intensive procedure, which makes it difficult to perform in all patient samples. PCR-based approaches can also be used for variant detection when required, provided the specific mutation has already been characterized. This technique can be used in selected situations when the identification of the variant is important, but may not be required for routine clinical decisions [27].

Overall, NAATs are currently the most sensitive tests available for diagnosing COVID-19. The turnaround time for lab-based NAATs is short enough to facilitate the diagnosis of individual patients as well as to allow the effective application of outbreak control measures. The turn-around time has been further shortened by POC tests. However, the false negativity rate can be high with POC tests, and they can be used to detect only current infections.

> The Omicron variant has been reported to carry a deletion in the S gene that can cause a negative signal for this target in PCR assays. This S-gene target failure has the potential to be used as a marker for the variant. Assays that use more than one target, as recommended, are unlikely to be affected.

Antigen detection tests

Tests detecting viral antigens in human biological specimens are an important category of direct investigations for the diagnosis of current COVID-19 infection. The antigens are usually viral proteins, which are otherwise not expected to be present in human specimens. While there are lab-based antigen tests, the large majority are used for rapid POC diagnosis. This is a major advantage over laboratory-based NAATs. However, the sensitivity is much lower than that of NAATs, as there is no amplification of the target analyte. Therefore, some institutions have evolved policies where positive antigen tests are considered true positive, whereas negative tests are confirmed by RT-PCR.

Most rapid antigen test (RAT) kits are based on a lateral flow assay platform that employs a nylon or nitrocellulose strip within a plastic cartridge. Application of a biological specimen within a well at one end of the cartridge results in movement of the target antigens through the strip by capillary action, during which they get bound to specific labeled antibodies, present within a "conjugate pad" of the strip. On moving further, the antigen–antibody complex gets immobilized at the "test line" and produces a visual signal due to the labeled antibodies, whereas the "control line" signals a functional test. In case the specimen does not contain the antigen, only the "control line" is visible. The turnaround time for most of the commercially available RAT kits is between 15 and 30 min, and all materials required to perform the test are generally provided with the kit. The SARS-CoV-2 N protein is the most commonly used target antigen [28,29].

Although the rapidity of results makes RATs useful for diagnosis in individual as well as public health situations, they have some limitations. These include lower sensitivity compared to RT-PCR. Additionally, the accuracy of RATs varies from one assay to the other, which makes it difficult to generalize information on RATs. A recently published Cochrane review estimated the sensitivity of rapid POC antigen tests to be 72.0% compared to RT-PCR, although the specificity was as high as 99.5% and 98.9% for symptomatic and asymptomatic individuals respectively [27]. The positive predictive value ranged from 84% to 90% for the two most sensitive assays performed in symptomatic subjects, for an assumed population prevalence of 5% [30]. Therefore, while RATs are good at identifying people who are not actually infected, there is a concern about missing true cases, which might need to be addressed by either repeating the test or performing an RT-PCR. For an individual, a positive test needs to be interpreted in the correct clinical context, that is, prevalence in the community and history of contact, and symptomatology. In view of these concerns, there are country- or region-specific guidelines with regard to the use of RATs.

There are various RATs currently available in the market, with variable performance characteristics. The WHO recommends a minimum sensitivity and specificity of at least 80% and 97%, respectively, in suspected COVID-19 cases, compared to an NAAT [31]. The test results are most reliable when the baseline population prevalence is at least 5%, that is, when there is ongoing community transmission [31]. It is also important to adhere to the manufacturer's instructions including the kit storage conditions (especially temperature) to ensure optimum performance. Most of the RAT kits are designed to test respiratory secretions or oral fluid/saliva. Similar to NAATs, antigen tests are most likely to be positive early in the course of the infection, that is, from 2 to 3 days before, to 5−7 days after, the onset of symptoms. There is also a positive correlation between the sensitivity of the test and the viral load, represented by the Ct value in various studies [30]. Self-testing with RAT is also possible; however, currently, there is no clear consensus about the accuracy of this practice [31].

In view of the constraints associated with the use of antigen tests, countries are expected to regulate the approval and marketing of brands and also, formulate guidelines for their interpretation, keeping in mind the local epidemiology and availability of human and nonhuman resources. The general principle is to consider RAT results as definitive if they are in line with the clinical suspicion and to repeat the test if they are not. Currently, the WHO considers the use of RATs helpful in the following situations [31]:

a. Symptomatic individuals: When symptomatic individuals meeting the case definition of suspected COVID-19 test positive on RAT, they should be considered infected and isolated and contact tracing should be done. For those who test negative, but a clinical suspicion persists, retesting with NAAT (if results can be available within 24 h) or repetition of RAT should be considered.

b. Asymptomatic individuals: In this group, testing with RAT should be limited to contacts of confirmed or probable cases and healthcare workers at risk. The need for confirmation of a positive test should be individualized depending upon the setting.

c. Outbreak investigation: Positive results from multiple individuals in a closed setting like schools or day-care centers could be highly suggestive of a COVID-19 outbreak and merits initiation of appropriate infection control measures.

To summarize, antigen tests score over NAATs in terms of ease of use, lower cost, and faster turnaround time. Specificity is comparable to that of NAATs. However, due to lower sensitivity and lower positive predictive value in low prevalence situations, they might need confirmation with a second test.

Viral culture

SARS-CoV-2 has been successfully isolated in cell lines and has been demonstrated to show cytopathic effects [32]. However, viral isolation is a time-consuming procedure that requires at least a biosafety level-3 (BSL-3) facility, along with highly trained staff. Therefore, it is not used for routine diagnosis and is mostly restricted to research purposes. The WHO and the CDC do not recommend using viral culture as a diagnostic procedure [8,33].

Indirect tests

This group includes some tests that are more readily available or accessible, hence could be considered as screening tools in some situations, to identify those who need to undergo definitive tests for COVID-19. Some of these tests may also be used to monitor people with the confirmed disease to make decisions regarding the need for hospital admission, transfer to an intensive care unit (ICU), and initiation of specific medical therapies. Some others may indicate past COVID-19 infection.

However, as expected, the results of such investigations are not specific to COVID-19 and require careful interpretation, keeping in mind the clinical setting, history of exposure (or contact), timing of onset of symptoms, and the probability of having a non-COVID-19 illness. The results of these tests cannot by themselves be taken as conclusive evidence of current infection. The situation might be slightly different for radiological investigations as described below.

Antibody tests

Tests detecting antibodies against SARS-CoV-2 antigens are considered indirect tests because these antibodies are produced by the body in response to the pathogen and not by the pathogen itself. Owing to the time lag between infection and the generation of antibodies, their estimation does not help to diagnose acute infection. However, it is the only method to identify individuals with past infection.

Knowledge about the immune response to SARS-CoV-2 is still evolving. However, knowledge of humoral responses to Middle East Respiratory Syndrome (MERS) and SARS-CoV-1 facilitates the interpretation of SARS-CoV-2 antibody tests. Current literature suggests that there is not much difference between the timing of IgM and IgG seroconversion to SARS-CoV-2; therefore, testing for IgM antibodies does not offer any real advantage. Most of the patients infected with the virus have been reported to develop detectable IgG levels by 2 weeks after symptom onset and almost all do so by 4 weeks. The antibodies are directed against the viral N and S proteins. Neutralizing antibodies directed specifically against the receptor binding domain (RBD) of the S1 subunit of viral S protein correlate with protection and were found in 90% of individuals by 4 weeks [34]. The duration of detectable antibodies after acute infection or vaccination is unclear but it is expected to be 1−2 years. There is some protection against reinfection, especially against symptomatic disease, but there is uncertainty about the duration of protection and the extent to which virus variants can evade it. The robustness of the immune response also varies depending upon the severity of illness, age, comorbidities, vaccination status, etc.; hence, these factors impact antibody detection and/or quantification [34−36].

There are predominantly two types of antibody assays in use: those detecting binding antibodies and those detecting neutralizing antibodies. Tests detecting binding antibodies use purified viral proteins and can be performed in BSL-2 facilities. They can be conducted using POC kits or lab-based methods. Tests detecting the presence of neutralizing antibodies need to be conducted in a laboratory, as it involves the in vitro determination of the ability to prevent SARS-CoV-2 infection. Antibody tests may detect either total immunoglobulin or specific components such as IgM, IgG, or IgA. Currently, there is no clinical significance of IgA assays for diagnosis. Antibody tests may also be qualitative, which report the result as positive or negative; semiquantitative, which give some estimate of the antibody levels; or quantitative, which provide the exact antibody titer [37].

Although antibody testing does not help with the diagnosis of acute infection for individual patient management, a rise in antibody titer over time could indicate a recent infection. Antibody testing has great value for epidemiological purposes, to retrospectively investigate the size of an outbreak, estimate the attack rate, etc. Seropositivity for SARS-CoV-2 can also be taken as evidence of previous infection, which fulfills one of the criteria for diagnosing multisystem inflammatory syndrome in children [38].

It cannot be overemphasized that antibody testing is currently not to be used to ascertain whether an individual is protected against SARS-CoV-2 as a result of vaccination or to assess the need for vaccination. This is because most of the current vaccines are supposed to induce neutralizing antibodies, specifically directed against the RBD of viral S protein or other epitopes depending upon the vaccine design. However, none of the currently approved serological tests have been validated to detect these specific antibodies that confer protection [8]. In addition, it is likely that there is some involvement of cell-mediated immunity also in protecting against SARS-CoV-2, which cannot be detected by antibody tests alone [39].

Radiological investigations

During the initial months of the pandemic, some studies (mostly from China) suggested that radiological investigations, especially computed tomography (CT) of the chest, were an efficient diagnostic method. The explanation offered was that chest CT is more sensitive than the gold standard RT-PCR during the early stages of the infection. At that time, RT-PCR was associated with high false negative rates, delaying timely implementation of infection control practices. Some studies suggested that CT scans could be repeated to monitor the clinical progression of the disease. An early version of a case definition for surveillance of COVID-19 by the Chinese Health Commission considered typical CT findings sufficient for a "clinical diagnosis" [40,41]. A consensus statement formulated by a number of groups focusing on pediatric COVID-19 in China used CT scan findings to classify clinical severity [42].

However, none of the imaging findings of SARS-CoV-2 pneumonia were specific enough to merit inclusion as a routine diagnostic test. The high cost, radiation exposure, and difficulty in performing scans in sick patients led the WHO to publish a rapid advice guidance in June 2020, recommending against using chest imaging to diagnose asymptomatic patients or even symptomatic patients, when timely RT-PCR testing was available [43]. However, the WHO suggested that chest imaging could be one of the investigations in symptomatic individuals when RT-PCR was not available, delayed or negative, despite a high index of suspicion. It also recommended against using imaging for making decisions about discharging hospitalized patients. In suspected or confirmed COVID-19 patients, chest imaging was sometimes included as an element of patient evaluation, to decide between home discharge and hospital admission for mild symptoms, and between regular ward and ICU admission, or to inform therapeutic management for moderate-to-severe symptoms. Most of these recommendations were based on expert opinion with low certainty of evidence [43].

With greater knowledge and experience gained over two years of the ongoing pandemic, the place of radiological investigations is now better understood.

Chest x-ray: X-ray is usually the first imaging modality in most healthcare settings. X-ray findings are expected to correlate with CT findings, although they can miss less severe abnormalities. Common x-ray findings in COVID-19 pneumonia include multifocal (patchy or confluent) peripheral lung opacities. There is usually involvement of more than one lobe, with a bilateral, lower lung distribution. Ground glass opacities (GGOs) seen on CT scans might be difficult to detect but accompanying reticular opacities can be seen more easily. Pleural effusion, cavitation, and pneumothorax are rare findings but have been reported. The x-ray findings usually become most evident between 6 and 12 days after the initial presentation [44].

Chest radiography is less resource-intensive, available in practically all settings and can be repeated more often. Radiation exposure is much less compared to CT scans. X-rays can be performed at the POC with portable equipment, avoiding the need to transport a sick or potentially contagious patient. However, the sensitivity of chest x-rays is lower than CT scans, as is the case for most conditions [43]. Therefore, chest X-rays are generally used for monitoring sick patients and during the follow-up [45].

CT scan: High-resolution CT (HRCT) is the radiological modality of choice in COVID-19 pneumonia. However, when pulmonary embolism is suspected, CT pulmonary angiography is required. Common HRCT findings include GGOs, which may have superimposed septal thickening (crazy paving pattern) and consolidation. These findings usually have a bilateral, multilobar distribution, with a predilection for subpleural, peripheral, and posterior regions. Based on the temporal evolution of lesions, 4 distinct stages have been described: early, progressive, peak, and absorption, with slightly variable durations. In the early stage, GGOs predominate. A relatively specific sign of COVID-19 pneumonia, better visualized on maximum intensity projection reconstructions, is the enlargement of the subsegmental vessels within the area of ground glass opacities. During the progressive stage, there is an increase in the extent and density of lesions and the appearance of consolidation with the typical distribution. Maximum radiological changes are seen in the peak or severe stage, when the characteristic signs, such as "reversed halo sign," "band-like opacities" and perilobular opacities appear, sometimes with traction bronchiectasis and bronchiolectatsis. This is followed by the absorption stage, during which there is clearing of the described lesions with variable persistence of GGOs, linear scarring, and residual consolidation. A decrease in the density of lesions, with more extensive involvement seen at this stage, is called "tinted" sign [45,46]. The CT severity score (CT-SS), based on the sum of scores for parenchymal opacification of 20 regions of the lung, each scored from 0 to 2, has been used as an objective way to identify severe disease [47].

Less frequent CT scan findings include air leak with pneumothorax, pneumomediastinum and subcutaneous emphysema, pleural effusion and thickening, cavitation, mediastinal lymphadenopathy, multifocal nodules, bronchial wall thickening, predominant anterior involvement, and unilateral or even unilobar involvement. Important radiological differentials that need to be considered in children include viral, bacterial or atypical pneumonia, cardiogenic pulmonary edema, aspiration pneumonia, and diffuse alveolar hemorrhage [46].

Similar to RT-PCR, CT also suffers from low sensitivity in the early stage of infection and the specificity is moderate to low. In areas with a low prevalence of COVID-19, the use of CT is expected to be associated with a high false positive rate, leading to wastage of healthcare resources. The downsides of CT scanning include the need for transporting patients, decontamination of the imaging room, and radiation exposure. This calls for judicious use of this investigation [45,46,48].

To address these issues and prevent indiscriminate use of chest CT during the pandemic, the CDC and the American College of Radiology recommended that CT should not be used for screening or diagnosing COVID-19 [49]. A more detailed multinational consensus statement on the role of chest imaging in patient management, considering different scenarios based on severity of symptoms, pretest probability, risk of disease progression, and resource constraints, has been published by the Fleischner Society [50]. There have been attempts to standardize the reporting language also. These include the CO-RADS scheme proposed by the Dutch Radiological Society and another system proposed by the Radiological Society of North America [51,52].

Ultrasonography: There is increasing interest in the use of ultrasonography (USG), especially bedside USG, for acute as well as chronic lung diseases. Important findings in COVID-19 include multiple separated or confluent B lines and subpleural consolidation. While none of these findings is pathognomonic of COVID-19, serial USG examinations can show the progression of the disease, with initial focal B lines gradually increasing in number and appearing at more sites and finally, coalescing to produce a consolidation pattern. Similarly, resolution of the disease can be seen, providing support for a decision to de-escalate therapy. The spatial density of B lines can also help with clinical management decisions such as prone positioning and titration of ventilator settings [45,53]. The lung USG findings in children have been reported to be similar to adults, in small studies [54].

The advantages of USG include its noninvasive, radiation-free nature, availability at the bedside, and repeatability if required. However, it is operator dependent and measures are required to ensure sterility of the equipment, and to shield the operator. The sensitivity is not high enough to allow its use as a standalone modality for diagnosis [48].

Other modalities: There has been some interest in using positron emission tomography combined with CT scan and also magnetic resonance imaging (MRI) in COVID-19. Both modalities offer the opportunity to assess extrapulmonary involvement, and MRI is a radiation-free option. However, availability is extremely limited and more experience is required to identify their role [45,48].

There is limited data on imaging findings of COVID-19 pneumonia in children. Chest X-ray and CT findings in children also lack the specificity required to make the initial diagnosis reliably. As infants and young children are more likely than adults to present with an acute respiratory syndrome of viral origin, this complicates the interpretation of findings. It has been suggested that GGOs due to SARS-CoV-2 in children are more localized, have lower attenuation, and have less lobular involvement, as compared to those in adults [55]. The WHO rapid advice guide suggests to prefer chest X-rays over CT in infants and young children [43].

Laboratory markers

Early in the course of the pandemic itself, it was realized that certain patterns in hematological parameters, inflammatory markers, and organ function

TABLE 5.1 Lab markers associated with COVID-19 infection and disease severity.

Haematological abnormalities	Biochemical derangements	Coagulation abnormalities	Inflammatory biomarkers
↑ Leukocyte count ↑ Neutrophil count ↓ Lymphocyte count ↓ Eosinophil count ↓ Platelet count ↓ Hemoglobin ↑ Neutrophil lymphocyte ratio	↓ Albumin ↑ Liver enzymes ↑ Total bilirubin ↑ BUN ↑ Creatinine ↑ CK, CK-MB ↑ LDH ↑ Myoglobin ↑ Cardiac troponin ↑ BNP	↑ Prothrombin time ↑ Activated partial thromboplastin time ↑ D-dimer	↑ ESR ↑ CRP ↑ Serum ferritin ↑ Procalcitonin ↑ SAA ↑ IL-2R, IL-6, IL-8, IL-10 ↑ TNF-α ↓ IFN-γ ↑ IP-10 and MCP

↑: Increased; ↓: decreased; *BNP*, brain natriuretic peptide; *BUN*, blood urea nitrogen; *CRP*, C-reactive protein; *CK*, creatine kinase; *ESR*, erythrocyte sedimentation rate; *LDH*, lactate dehydrogenase; *IFN*, interferon; *IL*, interleukin; *IP*, interferon-γ inducible protein; *MCP*, monocyte chemotactic protein; *SAA*, serum amyloid A; *TNF*, tumor necrosis factor.

indices were more frequently observed in patients diagnosed with COVID-19, especially those with more severe disease. These included lymphopenia, leukocytosis, raised CRP, and some others [56]. However, they are not specific to COVID-19 infection, as most of these markers reflect systemic inflammation and deranged organ function. However, systematic analysis of lab data showed that some of these have value in distinguishing between severe and nonsevere cases, and even predicting outcomes such as mortality [57–59]. Table 5.1 provides a nonexhaustive summary list of these lab markers.

These biomarkers have several limitations, in addition to their nonspecific nature. Some like interleukin levels might not be available at all places. Further, absolute cut-off values to distinguish between different groups of patients have not been established owing to significant heterogeneity in the patient population, disease and severity definitions, and timing of the test. Studies focusing only on the pediatric population have shown similar results [60,61].

Summary

Accurate diagnosis of COVID-19 infection is important to enable appropriate clinical decisions as well as initiation of infection control measures. It is important to realize that different investigations can provide different yet relevant information and need to be utilized judiciously. Diagnostic methods do not vary significantly between adults and children.

Direct tests demonstrate the presence of intact viral particles or their components. Nucleic acid amplification tests (NAATs) are the most sensitive of all investigations, and lab-based RT-PCR is currently the gold standard for diagnosis, despite the sensitivity being less than desirable. Other NAAT techniques are emerging and can support rapid, point-of-care testing. Accuracy depends upon the timing of testing and optimum collection, storage, and processing of respiratory samples. Antigen tests with shorter turnaround times are being used, predominantly in outbreak control scenarios. Their accuracy is less than that of RT-PCR and negative results might need confirmation. Use of viral culture is limited to research settings, on account of being technically challenging, requirement of BSL 3 facilities, and risk of exposure to lab personnel.

Indirect tests are based upon the identification of changes in the body, secondary to infection by SARS-CoV-2. Antibody testing, which is largely limited to IgG, can be used to support a retrospective diagnosis and is important for epidemiological purposes. Currently available antibody tests cannot be used to assess protection conferred by vaccines. Imaging is often done in children presenting with acute respiratory illness due to COVID-19 pneumonia; however, none of the imaging modalities including chest X-ray, CT, and ultrasound (USG) are sufficiently specific. Chest X-ray and USG can be used to monitor disease progression. Nonspecific derangements of hematological and biochemical parameters and some inflammatory markers have been observed in SARS-CoV-2 disease and could be helpful in monitoring severity.

Case studies

Case 1

A 2 year-old-boy presented with a history of choking while being fed peanuts, followed by bouts of cough and progressively increasing swelling over the left side of the face. On examination, he had tachypnea, intercostal retractions, hypoxia in room air, subcutaneous emphysema over the left side of the face and neck, rightward mediastinal shift, and reduced breath sounds over the left hemithorax. A clinical diagnosis of foreign body aspiration with air leak was considered. Chest X-ray showed a hyperinflated left lung with subcutaneous emphysema (Fig. 5.1).

The parents, who were feeding the child, had tested positive for COVID-19, five days before the event. They had been tested as they had a fever and upper respiratory tract symptoms for three days but were currently afebrile for three days. The child had been in contact with the parents but had not developed any symptoms.

The child was stabilized in an isolation room. The expected turnaround time for COVID-19 RT-PCR was around 12–24 h, but an urgent rigid

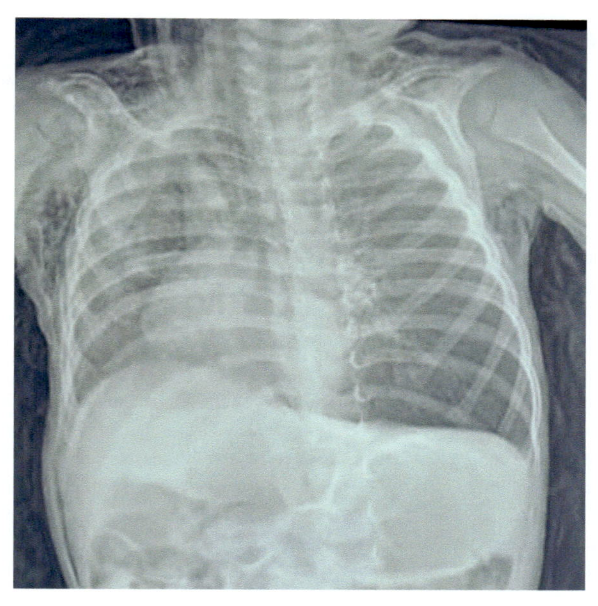

FIGURE 5.1 Chest X-ray showing hyperinflated left lung and subcutaneous emphysema.

bronchoscopy for foreign body removal was indicated. Hence, a nasopharyngeal swab was sent for Xpert Xpress SARS-CoV-2 test. A positive result was received within an hour, following which the patient was shifted to the operating room designated for COVID-19-positive patients, and the peanut was removed from the left lower lobe bronchus.

The child was then shifted to an isolation room in the pediatric ICU, where he was continued on mechanical ventilation for 12 h, and then extubated. The respiratory distress subsided and subcutaneous emphysema improved within 24 h. An RT-PCR was done after 3 days, which was reported to be negative, and he could be shifted to the general ward.

Case 2

A 14-year-old boy presented with fever, myalgia, and sore throat for 1 day. He reported visiting a friend's house 3 days previously. The friend later tested positive for COVID-19, on the same day.

As the child fulfilled the definition of a suspected case, a nasopharyngeal swab was taken and subjected to a rapid antigen test. The test was performed as per the kit instructions and read after 20 min. The result was negative.

However, as the child had developed symptoms and there was a high suspicion of COVID-19 infection, an RT-PCR was sent to the lab, while the patient was instructed to remain isolated at home and report if there were any troublesome symptoms. The RT-PCR report received after 24 h was positive.

The child became afebrile after 2 more days and did not develop any new symptoms.

References

[1] Zhou P, Yang X-L, Wang X-G, Hu B, Zhang L, Zhang W, et al. A pneumonia outbreak associated with a new coronavirus of probable bat origin. Nature March 2020;579(7798):270–3.

[2] Udugama B, Kadhiresan P, Kozlowski HN, Malekjahani A, Osborne M, Li VYC, et al. Diagnosing COVID-19: the disease and tools for detection. ACS Nano April 28, 2020;14(4):3822–35.

[3] Taleghani N, Taghipour F. Diagnosis of COVID-19 for controlling the pandemic: a review of the state-of-the-art. Biosens Bioelectron February 15, 2021;174:112830.

[4] Tosif S, Neeland MR, Sutton P, Licciardi PV, Sarkar S, Selva KJ, et al. Immune responses to SARS-CoV-2 in three children of parents with symptomatic COVID-19. Nat Commun November 11, 2020;11(1):5703.

[5] World Health Organization. COVID-19 disease in children and adolescents: scientific brief. September 29, 2021 [Internet]. WHO; [cited 2022 Jan 1]. Available from: https://www.who.int/publications/i/item/WHO-2019-nCoV-Sci_Brief-Children_and_adolescents-2021.1.

[6] Baggio S, L'Huillier AG, Yerly S, Bellon M, Wagner N, Rohr M, et al. Severe acute respiratory syndrome coronavirus 2 (SARS-CoV-2) viral load in the upper respiratory tract of children and adults with early acute coronavirus disease 2019 (COVID-19). Clin Infect Dis Off Publ Infect Dis Soc Am July 1, 2021;73(1):148–50.

[7] Madera S, Crawford E, Langelier C, Tran NK, Thornborrow E, Miller S, et al. Nasopharyngeal SARS-CoV-2 viral loads in young children do not differ significantly from those in older children and adults. Sci Rep February 4, 2021;11(1):3044.

[8] World Health Organization. Diagnostic testing for SARS-CoV-2: interim guidance. September 11, 2020 [Internet]. WHO; [cited 2022 Jan 20]. Available from: https://www.who.int/publications/i/item/diagnostic-testing-for-sars-cov-2.

[9] Centers for Disease Control and Prevention. Overview of testing for SARS-CoV-2 (COVID-19) [Internet]. CDC; 2021 [cited 2022 Jan 20]. Available from: https://www.cdc.gov/coronavirus/2019-ncov/hcp/testing-overview.html.

[10] Axell-House DB, Lavingia R, Rafferty M, Clark E, Amirian ES, Chiao EY. The estimation of diagnostic accuracy of tests for COVID-19: a scoping review. J Infect November 2020;81(5):681–97.

[11] Kostoulas P, Eusebi P, Hartnack S. Diagnostic accuracy estimates for COVID-19 real-time polymerase chain reaction and lateral flow immunoassay tests with Bayesian latent-class models. Am J Epidemiol August 1, 2021;190(8):1689–95.

[12] Pu R, Liu S, Ren X, Shi D, Ba Y, Huo Y, et al. The screening value of RT-LAMP and RT-PCR in the diagnosis of COVID-19: systematic review and meta-analysis. J Virol Methods February 2022;300:114392.

[13] Wikramaratna PS, Paton RS, Ghafari M, Lourenço J. Estimating the false-negative test probability of SARS-CoV-2 by RT-PCR. Euro Surveill Bull Eur Sur Mal Transm Eur Commun Dis Bull December 2020;25(50).

[14] Wang W, Xu Y, Gao R, Lu R, Han K, Wu G, et al. Detection of SARS-CoV-2 in different types of clinical specimens. JAMA May 12, 2020;323(18):1843–4.

[15] Lai CKC, Chen Z, Lui G, Ling L, Li T, Wong MCS, et al. Prospective study comparing deep throat saliva with other respiratory tract specimens in the diagnosis of novel coronavirus disease 2019. J Infect Dis October 13, 2020;222(10):1612–9.

[16] Moreira VM, Mascarenhas P, Machado V, Botelho J, Mendes JJ, Taveira N, et al. Diagnosis of SARS-Cov-2 infection by RT-PCR using specimens other than naso- and oropharyngeal

swabs: a systematic review and meta-analysis. Diagn Basel Switz February 21, 2021;11(2):363.

[17] Atieh MA, Guirguis M, Alsabeeha NHM, Cannon RD. The diagnostic accuracy of saliva testing for SARS-CoV-2: a systematic review and meta-analysis. Oral Dis June 3, 2021. https://doi.org/10.1111/odi.13934.

[18] Wang Y, Chen X, Wang F, Geng J, Liu B, Han F. Value of anal swabs for SARS-COV-2 detection: a literature review. Int J Med Sci 2021;18(11):2389–93.

[19] Chen Y, Chen L, Deng Q, Zhang G, Wu K, Ni L, et al. The presence of SARS-CoV-2 RNA in the feces of COVID-19 patients. J Med Virol July 2020;92(7):833–40.

[20] Sethuraman N, Jeremiah SS, Ryo A. Interpreting diagnostic tests for SARS-CoV-2. JAMA June 9, 2020;323(22):2249–51.

[21] Weiss A, Jellingsø M, Sommer MOA. Spatial and temporal dynamics of SARS-CoV-2 in COVID-19 patients: a systematic review and meta-analysis. EBioMedicine August 2020;58:102916.

[22] Rao SN, Manissero D, Steele VR, Pareja J. A systematic review of the clinical utility of cycle threshold values in the context of COVID-19. Infect Dis Ther September 2020;9(3):573–86.

[23] Shah VP, Farah WH, Hill JC, Hassett LC, Binnicker MJ, Yao JD, et al. Association between SARS-CoV-2 cycle threshold values and clinical outcomes in patients with COVID-19: a systematic review and meta-analysis. Open Forum Infect Dis September 2021;8(9):ofab453.

[24] Rabaan AA, Tirupathi R, Sule AA, Aldali J, Mutair AA, Alhumaid S, et al. Viral dynamics and real-time RT-PCR Ct values correlation with disease severity in COVID-19. Diagn Basel Switz June 15, 2021;11(6):1091.

[25] Engelmann I, Alidjinou EK, Ogiez J, Pagneux Q, Miloudi S, Benhalima I, et al. Pre-analytical issues and cycle threshold values in SARS-CoV-2 real-time RT-PCR testing: should test results include these? ACS Omega March 16, 2021;6(10):6528–36.

[26] Yu CY, Chan KG, Yean CY, Ang GY. Nucleic acid-based diagnostic tests for the detection SARS-CoV-2: an update. Diagn Basel Switz January 1, 2021;11(1):53.

[27] Thomas E, Delabat S, Carattini YL, Andrews DM. SARS-CoV-2 and variant diagnostic testing approaches in the United States. Viruses December 13, 2021;13(12):2492.

[28] S Pavia C, Plummer M. The evolution of rapid antigen detection systems and their application for COVID-19 and other serious respiratory infectious diseases. J Microbiol Immunol Infect Wei Mian Yu Gan Ran Za Zhi October 2021;54(5):776–86.

[29] Mahmoudinobar F, Britton D, Montclare JK. Protein-based lateral flow assays for COVID-19 detection. Protein Eng Des Sel PEDS February 15, 2021;34:gzab010.

[30] Dinnes J, Deeks JJ, Berhane S, Taylor M, Adriano A, Davenport C, et al. Rapid, point-of-care antigen and molecular-based tests for diagnosis of SARS-CoV-2 infection. Cochrane Database Syst Rev March 24, 2021;3:CD013705.

[31] World Health Organization. Antigen-detection in the diagnosis of SARS-CoV-2 infection: Interim guidance. October 6, 2021 [Internet]. WHO; [cited 2022 Jan 22]. Available from: https://www.who.int/publications/i/item/antigen-detection-in-the-diagnosis-of-sars-cov-2infection-using-rapid-immunoassays.

[32] Chu H, Chan JF-W, Yuen TT-T, Shuai H, Yuan S, Wang Y, et al. Comparative tropism, replication kinetics, and cell damage profiling of SARS-CoV-2 and SARS-CoV with implications for clinical manifestations, transmissibility, and laboratory studies of COVID-19: an observational study. Lancet Microbe May 2020;1(1):e14–23.

[33] Centers for Disease Control and Prevention. SARS-CoV-2 viral culturing at CDC [Internet]. CDC; 2020 [cited 2022 Jan 21]. Available from: https://www.cdc.gov/coronavirus/2019-ncov/lab/grows-virus-cell-culture.html.

[34] O Murchu E, Byrne P, Walsh KA, Carty PG, Connolly M, De Gascun C, et al. Immune response following infection with SARS-CoV-2 and other coronaviruses: a rapid review. Rev Med Virol March 2021;31(2):e2162.
[35] Baumgarth N, Nikolich-Žugich J, Lee FE-H, Bhattacharya D. Antibody responses to SARS-CoV-2: let's stick to known knowns. J Immunol Baltim Md 1950 November 1, 2020;205(9):2342−50.
[36] World Health Organization. COVID-19 natural immunity: scientific brief. May 10, 2021 [Internet]. WHO; [cited 2022 Jan 24]. Available from: https://www.who.int/publications/i/item/WHO-2019-nCoV-Sci_Brief-Natural_immunity-2021.1.
[37] Centers for Disease Control and Prevention. Interim guidelines for COVID-19 antibody testing [Internet]. CDC; 2021 [cited 2022 Jan 25]. Available from: https://www.cdc.gov/coronavirus/2019-ncov/lab/resources/antibody-tests-guidelines.html.
[38] Feldstein LR, Rose EB, Horwitz SM, Collins JP, Newhams MM, Son MBF, et al. Multisystem inflammatory syndrome in U.S. Children and adolescents. N Engl J Med July 23, 2020;383(4):334−46.
[39] Sadarangani M, Marchant A, Kollmann TR. Immunological mechanisms of vaccine-induced protection against COVID-19 in humans. Nat Rev Immunol August 2021;21(8):475−84.
[40] Zu ZY, Jiang MD, Xu PP, Chen W, Ni QQ, Lu GM, et al. Coronavirus disease 2019 (COVID-19): a perspective from China. Radiology August 2020;296(2):E15−25.
[41] Yang W, Sirajuddin A, Zhang X, Liu G, Teng Z, Zhao S, et al. The role of imaging in 2019 novel coronavirus pneumonia (COVID-19). Eur Radiol September 2020;30(9):4874−82.
[42] Shen K-L, Yang Y-H, Jiang R-M, Wang T-Y, Zhao D-C, Jiang Y, et al. Updated diagnosis, treatment and prevention of COVID-19 in children: experts' consensus statement (condensed version of the second edition). World J Pediatr WJP June 2020;16(3):232−9.
[43] World Health Organization. Use of chest imaging in COVID-19: a rapid advice guide. June 11, 2020 [Internet]. WHO; [cited 2022 Jan 27]. Available from: https://www.who.int/publications/i/item/use-of-chest-imaging-in-covid-19.
[44] Jacobi A, Chung M, Bernheim A, Eber C. Portable chest X-ray in coronavirus disease-19 (COVID-19): a pictorial review. Clin Imag August 2020;64:35−42.
[45] Churruca M, Martínez-Besteiro E, Couñago F, Landete P. COVID-19 pneumonia: a review of typical radiological characteristics. World J Radiol October 28, 2021;13(10):327−43.
[46] Larici AR, Cicchetti G, Marano R, Merlino B, Elia L, Calandriello L, et al. Multimodality imaging of COVID-19 pneumonia: from diagnosis to follow-up. A comprehensive review. Eur J Radiol October 2020;131:109217.
[47] Yang R, Li X, Liu H, Zhen Y, Zhang X, Xiong Q, et al. Chest CT severity score: an imaging tool for assessing severe COVID-19. Radiol Cardiothorac Imaging April 2020;2(2):e200047.
[48] de Carvalho LS, da Silva Júnior RT, Oliveira BVS, de Miranda YS, Rebouças NLF, Loureiro MS, et al. Highlighting COVID-19: what the imaging exams show about the disease. World J Radiol May 28, 2021;13(5):122−36.
[49] Centers for Disease Control and Prevention. Interim clinical guidance for management of patients with confirmed coronavirus disease (COVID-19) [Internet]. CDC; 2021 [cited 2022 Jan 29]. Available from: https://www.cdc.gov/coronavirus/2019-ncov/hcp/clinical-guidance-management-patients.html.
[50] Rubin GD, Ryerson CJ, Haramati LB, Sverzellati N, Kanne JP, Raoof S, et al. The role of chest imaging in patient management during the COVID-19 pandemic: a multinational consensus statement from the Fleischner society. Radiology July 2020;296(1):172−80.

[51] Prokop M, van Everdingen W, van Rees Vellinga T, Quarles van Ufford H, Stöger L, Beenen L, et al. CO-RADS: a categorical CT assessment scheme for patients suspected of having COVID-19-definition and evaluation. Radiology August 2020;296(2):E97–104.
[52] Simpson S, Kay FU, Abbara S, Bhalla S, Chung JH, Chung M, et al. Radiological society of North America expert consensus document on reporting chest CT findings related to COVID-19: endorsed by the society of thoracic Radiology, the American College of Radiology, and RSNA. Radiol Cardiothorac Imaging April 2020;2(2):e200152.
[53] Kameda T, Mizuma Y, Taniguchi H, Fujita M, Taniguchi N. Point-of-care lung ultrasound for the assessment of pneumonia: a narrative review in the COVID-19 era. J Med Ultrason 2001 January 2021;48(1):31–43.
[54] Musolino AM, Supino MC, Buonsenso D, Ferro V, Valentini P, Magistrelli A, et al. Lung ultrasound in children with COVID-19: preliminary findings. Ultrasound Med Biol August 2020;46(8):2094–8.
[55] Duan Y-N, Zhu Y-Q, Tang L-L, Qin J. CT features of novel coronavirus pneumonia (COVID-19) in children. Eur Radiol August 2020;30(8):4427–33.
[56] Frater JL, Zini G, d'Onofrio G, Rogers HJ. COVID-19 and the clinical hematology laboratory. Int J Lab Hematol June 2020;42(Suppl. 1):11–8.
[57] Henry BM, de Oliveira MHS, Benoit S, Plebani M, Lippi G. Hematologic, biochemical and immune biomarker abnormalities associated with severe illness and mortality in coronavirus disease 2019 (COVID-19): a meta-analysis. Clin Chem Lab Med June 25, 2020;58(7):1021–8.
[58] Tjendra Y, Al Mana AF, Espejo AP, Akgun Y, Millan NC, Gomez-Fernandez C, et al. Predicting disease severity and outcome in COVID-19 patients: a review of multiple biomarkers. Arch Pathol Lab Med December 1, 2020;144(12):1465–74.
[59] Al-Saadi EAKD, Abdulnabi MA. Hematological changes associated with COVID-19 infection. J Clin Lab Anal November 2021;16:e24064.
[60] Zimmermann P, Curtis N. COVID-19 in children, pregnancy and neonates: a review of epidemiologic and clinical features. Pediatr Infect Dis J June 2020;39(6):469–77.
[61] Cui X, Zhang T, Zheng J, Zhang J, Si P, Xu Y, et al. Children with coronavirus disease 2019: a review of demographic, clinical, laboratory, and imaging features in pediatric patients. J Med Virol September 2020;92(9):1501–10.

Chapter 6

Management of acute COVID-19 in the pediatric population and role of antimicrobial therapy

H.E. Groves[1], U. Allen[2] and S.K. Morris[2]
[1]*Wellcome-Wolfson Institute for Experimental Medicine, Queens University, Belfast, Northern Ireland, United Kingdom;* [2]*Division of Infectious Diseases, The Hospital for Sick Children, University of Toronto, Toronto, ON, Canada*

Introduction

Understanding of the management of acute COVID-19 in children continues to evolve at a rapid pace. Severe disease and death with COVID-19 have been consistently demonstrated to occur relatively infrequently in children and young adults compared to older adults [1−3]. True values for hospitalization and mortality rates in children with COVID-19 infection are difficult to determine and regional variability may exist, with evidence suggesting that lower- and middle-income countries may bear a greater burden of pediatric COVID-19 mortality compared to higher-income countries [4]. Awareness of the risk of severe COVID-19 disease in children underpins appropriate clinical management, and the reader should clearly understand that the vast majority of children with acute COVID-19 will improve with supportive care alone [5]. An additional challenge posed by the relatively smaller number of severely affected children with acute COVID-19 is the ability to perform controlled clinical trials in the pediatric age group. As of January 2022, no pediatric-specific data for treatment is yet available from randomized controlled clinical trials. Accordingly, much of our current knowledge on the use of targeted therapies for children with COVID-19 is based on limited observational studies and extrapolation of data from research conducted primarily in the adult population.

Living guidance on the management of children with COVID-19 has been issued by a number of organizations including the World Health Organization (WHO) [6]. Publicly available country-specific guidance has also been issued

Clinical Management of Pediatric COVID-19. https://doi.org/10.1016/B978-0-323-95059-6.00006-1

by the United Kingdom (UK) Royal College of Pediatrics and Child Health (RCPCH), the Italian Society of Pediatric Infectious Diseases, and the United States National Institutes of Health (NIH) [7–9]. In this chapter, we address the evidence for the initial management of pediatric patients with acute COVID-19 with a particular focus on the role of targeted therapeutics. It should be noted that due to the rapid evolution of evidence, regularly updated local, national and regional guidance documents should be consulted when determining treatment of pediatric patients with acute COVID-19. The approach to infection control measures, investigations, management of complications of COVID-19, postinfectious SARS-CoV-2 infections, and long COVID will be covered elsewhere in this book.

Clinical application: management of children with acute COVID-19

General management considerations

For pediatric patients with acute COVID-19, supportive care and management of complications remain the mainstay of management [8,9]. Careful clinical assessment is vital and includes consideration that positive testing for COVID-19 may also represent an incidental finding in a child presenting with a separate illness [10]. Additionally, a positive PCR test for SARS-CoV-2 may reflect recent, but not acute infection and furthermore, serological evidence of SARS-CoV-2 antibodies should never be used to diagnose acute SARS-CoV-2 infection.

Use of specific treatments, including antiviral and immunomodulatory therapies, should be made as part of a multidisciplinary team discussion including, but not limited to, teams with expertise in infectious diseases, critical care, rheumatology, and respiratory disciplines. We support the view that as far as possible, investigational therapies for the treatment of COVID-19 in children should be only given in the context of clinical treatment trials [9,11,12]. All pediatric patients presenting with acute COVID-19 should be stratified according to disease severity and consideration given to potential risk factors for severe disease.

Criteria for determining the severity of COVID-19 in children

Variation exists in the literature regarding the definition of disease severity in children with COVID-19. Early in the pandemic, definitions of severity in children infected with SARS-CoV-2 included radiological criteria [13]. However, as the pandemic has progressed, clinically based classifications are now more favored [14]. Classifications in adults typically include consideration of the level of respiratory support required and/or presence of end-organ dysfunction as key components. The Infectious Diseases Society of America, for example, classifies adult cases of COVID-19 as mild-to-moderate/non-severe illness (no need for supplemental oxygen); severe illness (oxygen saturations $\leq$94% on room air) and critical illness (on mechanical ventilation or extracorporeal membrane oxygenation (ECMO) or sepsis/septic shock) [15]. This differs slightly from the severity definitions used by the WHO, which suggests a threshold of oxygen saturation <90% on room air to define severe COVID-19 [6].

An early clinical criterion for grading COVID-19 severity in children was proposed by Dong et al. in their report on the characteristics of COVID-19 in pediatric patients in China [16]. This classification included the use of radiological imaging findings in definitions or severity (see Table 6.1). Since this report, several updated classifications have been suggested by a number of groups including the WHO, the UK RCPCH, the Italian Society of Pediatric Infectious Diseases, and a multicenter panel of pediatric infectious diseases physicians and pharmacists from 18 geographically diverse North American institutions [6,7,9,12]. Similar to adult COVID-19 severity grading, each of these newer classifications is based on clinical criteria with a focus on respiratory support requirements. However, differences between classifications in the level of respiratory support used to delineate moderate, severe, and critical diseases exist as detailed in Table 6.1.

As the COVID-19 pandemic progresses with the emergence of new variants of concern on a frequent basis, our understanding of pediatric disease severity and its classification continues to evolve, with notable variation in the likelihood of severe disease observed between SARS-CoV-2 variants. Clinical judgment should be used on a case-by-case basis to determine clinical severity, taking into consideration institution-specific or national guidelines for classification where available. In general terms, disease severity is considered mild to moderate where no or minimal additional oxygen support is required. Where significant respiratory distress and increasing oxygen requirement occur, the disease is considered to be severe, with acute respiratory distress syndrome (ARDS), multiorgan failure, and need for invasive mechanical ventilation or greater support being considered critical disease.

Risk factors for severe COVID-19 in children

In children, respiratory viral infections with a respiratory syncytial virus (RSV) and influenza have well-established host risk factors for severe disease. Risk factors for severe RSV disease, for example, include prematurity, chronic lung disease such as bronchopulmonary dysplasia, congenital heart disease, trisomy 21, and the presence of underlying immunodeficiency, neurological or hemato-oncological disease [18–24]. Similarly, risk factors for severe influenza virus infection in children include young age, prematurity, presence of severe underlying medical conditions, and immunodeficiency [25–27]. Piroth et al. compared risk factors for severe disease between influenza and

TABLE 6.1 Classifications of COVID-19 Severity in Children. Summary of criteria used for classifying pediatric COVID-19 severity as detailed in listed available pediatric classifications as of the date 1st January 2022. Please note that assessment of disease severity may differ by age and adjustments in assessment may be required for very young infants and neonates.

Disease severity	Mild to moderate/ nonsevere disease	Severe disease	Critical disease
Dong et al. [16]	• **Mild:** symptoms of acute upper respiratory tract infection (fever, fatigue, myalgia, cough, sore throat, runny nose, sneezing) • **Moderate:** pneumonia, frequent fever, and cough; some may have wheezing but no obvious hypoxemia. Some cases may have no clinical signs or symptoms but chest computed tomography shows lung lesions	• Respiratory symptoms, such as fever and cough, may be gastrointestinal symptoms, disease progresses at ~1 week, and dyspnea with oxygen saturations <92%	• Acute respiratory distress syndrome or respiratory failure may also have shock, encephalopathy, myocardial injury or heart failure, coagulation dysfunction, and acute kidney injury
Italian Society of Pediatric Infectious diseases [7]	• **Mild:** fever and/or asthenia with upper respiratory signs, not requiring supplemental oxygen • **Moderate:** respiratory signs/ symptoms (such as cough, mild distress with polypnea) requiring supplemental oxygen with nasal cannulas or venturi system ± fever, difficulty in feeding, signs of dehydration	• Respiratory signs/symptoms (tachypnea, labored breathing) requiring supplemental oxygen with high flow nasal cannulas or noninvasive ventilation ± fever, systemic signs of worsening (lethargy, inability to feed/drink)	• Acute Respiratory Distress Syndrome (ARDS)**OR** • Respiratory involvement requiring mechanical ventilation or extracorporeal membrane oxygenation

Multicenter guidance, Chiotos et al. [12]	• No new or increased requirement for supplemental oxygen	• New or increase from baseline supplemental oxygen requirement **without** the need for new or increase in baseline noninvasive/invasive mechanical ventilation[a]	• New or increased requirement for invasive or noninvasive mechanical ventilation,[a] sepsis, or multiorgan failure **OR** • Rapidly worsening clinical trajectory that does not yet meet these criteria.
RCPCH [9]	• No oxygen requirement • Mild upper airway infection	• Mild to moderate ARDS[b]: **1.** Unventilated requiring FiO_2 >40% to maintain saturation 88%–97% **2.** Ventilation: - Oxygenation index: $4 \leq 16$ - Oxygenation saturation index: $5 \leq 12.3$	• Severe ARDS[b] - Oxygenation index: ≥ 16 - Oxygenation saturation index: ≥ 12.3 • Septic shock • Altered consciousness • Multiorgan failure
WHO [6]	• Absence of signs of severe or critical disease	• Oxygen saturation <90% on room air • Signs of severe respiratory distress including very severe chest wall in-drawing, grunting, central cyanosis, or presence of any other general danger signs (inability to breastfeed or drink, lethargy or reduced level of consciousness, convulsions) in addition to the signs of pneumonia	• Defined by the criteria for acute respiratory distress syndrome (ARDS), sepsis, septic shock, or other conditions that would normally require the provision of life-sustaining therapies such as mechanical ventilation (invasive or noninvasive) or vasopressor therapy

[a]*Noninvasive mechanical ventilation includes high-flow nasal cannula, continuous positive airway pressure (CPAP) and bilevel positive pressure (BiPAP).*
[b]*ARDS as defined by the PARD criteria: Paediatric Acute Lung Injury Consensus Conference Group [17].*

COVID-19 in hospitalized patients and found obesity, diabetes, hypertension, and dyslipidaemia were more commonly associated with severe COVID-19 than influenza, whereas those with severe influenza more frequently had heart failure, chronic respiratory disease, cirrhosis, and anemia [28]. The authors also noted the proportion of pediatric patients hospitalized was less for COVID-19 than for influenza.

Specific risk factors for severe and critical COVID-19 in the pediatric population remain to be clearly defined. As for RSV and influenza, young age and the presence of comorbidities such as chronic lung disease and neurological disease have been associated with an increased risk of hospitalization. One early large study by Dong et al. noted that over 60% of severe and critical cases of COVID-19 in children occurred in those aged 5 years or less [16]. In a cross-sectional study of 46 North American Pediatric Intensive Care Units (PICUs) with 48 children admitted secondary to COVID-19, 83% were noted to have significant preexisting comorbidities [29]. Of these, the most common reported comorbidity was children considered to be medically complex (long-term dependence on technological support, developmental delay, and genetic anomalies), followed by those with immune suppression or malignancy and those with obesity. Similar to this work, a more recent review of pediatric hospital admission with acute COVID-19 in Canada showed a number of factors associated with severe disease, including the presence of comorbid conditions such as chronic neurological conditions, chronic lung disease other than asthma, as well as obesity [30]. A large study of children hospitalized with acute COVID-19 in the United Kingdom found that critical care admission was more likely in neonates, black ethnicity, those born prematurely, and children with known comorbidities [31]. Additional limited observational evidence has also linked chronic cardiac disease and sickle cell disease to increased risk of hospitalization or severe COVID-19 in children and adolescents; however, these conditions have not been as consistently demonstrated to be risk factors for severe COVID-19 as the other factors listed earlier [32].

In relation to mortality data in children with SARS-CoV-2 infection, a review by the Centers for Disease Control and Prevention in the United States found more deaths reported in children aged 10–20 years compared to those under 10 years old, with most deaths occurring in Hispanic, non-Hispanic Black and non-Hispanic American Indian/Alaskan Native persons [33].

Extrapolating from these observational data and adult data, as well as from risk factors for severe disease in children with other human coronavirus infections, it might be reasonable to consider that children with significant comorbidities, such as chronic lung disease, genetic abnormalities, complex neurological disease or epilepsy, and diabetes mellitus as well as those with obesity may be at increased risk of severe infection [34]. In contrast to the evidence for increased risk of severe RSV and influenza disease in immunocompromised children, the evidence for severe COVID-19 in severely immunocompromised children is more limited. This may be in part due to smaller numbers of children with these conditions who have had COVID-19. Increased severity of SARS-CoV-2 infection has not been clearly described in pediatric solid organ transplant recipients or in children on chronic immunosuppressive therapy for renal, rheumatological, or gastrointestinal conditions [35,36]. In patients with hematological malignancies, adult data do suggest higher mortality due to COVID-19 in comparison with the general population; however, hematological malignancy as a risk factor for increased disease severity has not been consistently noted in the pediatric population [35,37]. It is also important to consider that for some conditions, the reason for increased hospitalization in children presenting with acute COVID-19 may not necessarily be due to clinically more severe disease, but might be explained by protocols recommending hospitalization for these conditions in the case of fever or respiratory symptoms [32]. Furthermore, there is increasing concern that many immunocompromised individuals may have less effective responses to COVID-19 vaccination and thus may not derive the same protection against severe disease as those who are immunocompetent. Accordingly, at present, the absolute risk for severe or critical disease in a child presenting with acute COVID-19 for any of the conditions listed earlier remains difficult to quantify.

Therapeutic management approach to hospitalized children with COVID-19

Provision of supportive care measures

In pediatric patients with acute COVID-19, management should include consideration of any underlying medical conditions such as asthma, neurological disorders, complex medical conditions, and immunosuppressive status. Management of these underlying comorbidities should follow standard practice, in consultation with clinicians who have expertise in the management of the child's underlying condition. Hospital admission is indicated for children with COVID-19 presenting with severe or critical illness. In those with nonsevere or mild illness, hospital admission may also be indicated if risk factors for severe disease as discussed above are present.

Supportive measures for children with COVID-19 should include consideration of a number of areas:

- Fluid and electrolyte management

Box 1. Fluid and electrolyte management is an important part of the general care of pediatric patients with acute COVID-19 as is the case with other infections in children, especially those who are very young.

Appropriate fluid support should follow regional and national guidance for pediatric patients such as that provided by the American Academy of Paediatrics [38] or the National Institute for Health and Care Excellence [39]. Consideration should be given to the potential for acute kidney injury as a result of the viral infection and monitoring may be required. In severe and critical cases fluid restriction may be required due to the risk of developing ARDS and expert advice from critical care and infectious diseases teams should be sought early [9].

- Respiratory support

Box 2. Escalation of respiratory support is an important aspect of the care of pediatric patients with acute COVID-19. As is the case for other causes of respiratory compromise in children, guidelines recommend use of noninvasive and invasive ventilation methods, ECMO and consideration of prone positioning should be directed by teams with relevant expertise.

Although most children with acute COVID-19 are unlikely to develop respiratory failure, the need for respiratory support should be regularly assessed as changes in respiratory status can occur rapidly [9]. UK and Australian guidelines suggest initial use of low flow nasal cannula oxygen in hypoxic children with acute COVID-19 with escalation to high flow nasal cannulae oxygen support on a case-by-case basis and early use of CPAP and other noninvasive ventilation to reduce the need for invasive mechanical ventilation [9,40]. PICU-specific guidance is also available on the management of the pediatric airway and indications for escalation to intubation in the context of COVID-19 (ref: https://icmanaesthesiacovid-19.org/covid-19-paediatric-airway-management-principles).

For the management of adults with COVID-19, there are a large number of international guidelines available outlining the use of noninvasive respiratory support and approaches to acute respiratory failure, including global guidance from the Surviving Sepsis Campaign and the WHO [6,41]. Significant variability exists in practice internationally due to an absence of established evidence [42] and Information on the specific management of ARDS in pediatric cases of COVID-19 is even more limited. Likewise, adult studies suggest mortality benefits with the use of prone positioning strategies in critically ill patients with ARDS secondary to COVID-19 and there is emerging evidence of benefits to its use in awake nonintubated patients [43–45]. However, while many guidelines recommend the use of proning in children with severe COVID-19, robust evidence on its use in this population is scarce [46]. Additional work to evaluate the most effective respiratory support mechanisms in children is urgently needed. In the meantime, in general, the principles of management of pediatric ARDS secondary to COVID-19 are likely to be aligned with those of the adult population, particularly in older children. However, there are key differences between pediatric and adult physiology as well as differences in the management of ARDS to consider with respect to the pediatric population [47]. Accordingly, decisions regarding noninvasive and subsequent escalation to invasive respiratory support including mechanical ventilation and use of ECMO should be made on an individual patient basis with involvement of respiratory and critical care teams when needed. It is also important to ensure the implementation of appropriate measures for staff protection and protective equipment use when providing additional respiratory support.

- Antipyretic and nonsteroidal antiinflammatory drug use

> Box 3. The use of nonsteroidal antiinflammatory drugs (NSAIDs) and other antipyretic medications may be considered as part of supportive care in children and adolescents with acute COVID-19 and, as for other infections, guidance exists regarding their appropriate use.

Early in the pandemic, there were initial concerns that the use of NSAID therapy might increase the risk of complications in patients with acute COVID-19 [48,49]. However, these concerns have not been supported by observational data and a meta-analysis of 11 observational studies demonstrated NSAIDs exposure did not increase the severity of COVID-19 disease [50]. Thus, most guidelines suggest the management of fever in children with COVID-19 should follow standard guidance as for other infections. Some guidelines suggest acetaminophen may be the preferred first line given the potential for NSAIDs to worsen acute kidney injury in children with poor fluid intake [9].

- Management of COVID-19-related coagulopathy

> Box 4. Coagulopathy in adult patients with acute COVID-19 has been reported and guidelines developed from emerging evidence direct the level of anticoagulation therapy to use in this context. In children, the risk is less clear and, due to limited evidence, guidelines differ with regard to the use of anticoagulation prophylaxis during acute COVID-19.

As the pandemic has evolved, reports from different centers have shown a high prevalence of coagulopathy in adults hospitalized with COVID-19, particularly in critically ill patients [51]. In the absence of an acute venous thromboembolism (VTE), hospitalized adults in a noncritically ill state seem to benefit from therapeutic-intensity, full dose, anticoagulation prophylaxis whereas critically ill patients benefit from prophylactic-intensity anticoagulation [52]. In children, the overall absolute thrombotic risk is not well defined and is likely to be lower than that of adults. Two multicenter cohort studies have indicated an incidence of thrombotic complications including deep vein thrombosis (DVT), pulmonary embolism (PE), and cerebral sinovenous thrombosis, ranging from 1.1% to 2.1% in hospitalized children with COVID-19 [53,54]. Thrombotic complications occurred more commonly in children aged over 12 years and were associated with severe or critical disease. Both studies identified a number of factors associated with increased risk of thrombosis including obesity, malignancy, and presence of central venous catheter.

While the risk for thrombotic events is higher in patients who develop multisystem inflammatory syndrome in children (MIS-C) [53], interventions to reduce the risk of venous thromboembolism (VTE) may be warranted for children hospitalized with acute COVID-19, especially in children over 12 years old or those with risk factors. Adult guidance on antithrombotic therapy in patients with COVID-19 has been published by several societies and governing bodies, including the NIH, the British Thoracic Society, and the American Society of Hematology [52,55,56]. These guidelines do not provide specific guidance on VTE prophylaxis in hospitalized children with COVID-19 beyond recommending the use of standard procedures as for other hospitalized children. The International Society on Thrombosis and Haemostasis [57] has issued a consensus-based clinical guidance suggesting prophylaxis intensity anticoagulation for children admitted with either acute COVID-19 or MIS-C, particularly if accompanied by additional thrombosis risk factors or by significant D-dimer level elevation. However, other pediatric publications have been less liberal, suggesting intervention only in adolescents with accompanying risk factors [58,59].

Investigations such as Doppler ultrasound should be urgently performed in patients with clinical signs of DVT, and PE should be considered in patients with sudden worsening of respiratory parameters out of keeping with other markers of COVID-19 disease severity [55]. The diagnostic value of D-dimer in pediatric DVT is limited and levels have been shown to have poor discriminative and predictive ability for DVT [60]. Furthermore, the interpretation of D-dimer results and other coagulation parameters in the context of COVID-19 infection is challenging given that elevated D-dimer and fibrinogen levels occur with severe COVID-19 [55]. Furthermore, in pediatric patients presenting with acute COVID-19 who develop confirmed VTE, no specific guidelines for management exist. Standard guidance for the treatment of pediatric VTE, such as the American Society of Hematology 2018 Guidelines [61], is available, and early discussion with a pediatric hematologist may be beneficial.

- Management of concomitant bacterial and fungal infection

Box 5. Treatment of bacterial and fungal infection in children and adolescents presenting with acute COVID-19 is based on the clinical scenarios as directed by local guidelines, with input from infectious diseases specialists where appropriate.

Bacterial coinfection and secondary infections are relatively infrequent occurrences in hospitalized pediatric patients with acute COVID-19 [62–64]. In children, there is significant variation by region on reported presence of bacterial coinfection with some studies suggesting lower incidence ranging from 0% to 4% and other reports finding up to 20% incidence [62,65]. Although challenges in differentiating upper airway colonization from true infection may lead to an overestimation of secondary infections in some studies and in addition, variation in secondary bacterial infection rates may also reflect differences in baseline population health characteristics and diagnostic resource availability. Use and choice of antibiotics should be directed by local policy for empiric antibiotics and based on standard practice for determining the presence of bacterial infection [9].

Several studies have highlighted that adult patients with COVID-19 requiring intensive care unit (ICU) level care are at risk of developing secondary fungal infections. The European Confederation of Medical Mycology and International Society for Human and Animal Mycology (ECMM/ISHAM) criteria have published definition criteria as well as management guidance for COVID-19-associated pulmonary aspergillosis (CAPA) in adults [66]. Evidence from the French Multicentre MYVCOVID study found that 15% of adult patients with severe COVID-19 in the ICU fulfilled ECMM/ISHAM CAPA criteria. Additional fungal infections with mucormycosis, invasive candidiasis, and *Pneumocystis jirovecii* have also been reported [67]. In children, fungal infection in association with COVID-19 has only been described in rare case reports, including rhino-orbito-cerebral mucormycosis in two pediatric patients with type 1 diabetes mellitus, which is a known independent risk factor for mucormycosis [68]. While rare, fungal infection in children with severe or critical COVID-19 should be considered in patients with compatible clinical syndromes given that early diagnosis and treatment are key to improving outcomes.

- Use of bronchodilators for asthma exacerbation

Box 6. There is no evidence to suggest that acute asthma exacerbation in children and adolescents who are positive for SARS-CoV-2 should be managed differently from standard practice for the treatment of asthma exacerbations.

In children presenting with acute wheeze or asthma exacerbation in the context of a positive SARS-CoV-2 test, guidelines suggest treatment with glucocorticoid and bronchodilators should follow standard guidance for the management of acute asthma or wheeze [69]. Consideration may be given to the use of inhaled rather than nebulized therapy if appropriate to reduce the risk of virus aerosolization [59].

Role of specific therapies against COVID-19

The optimal approach to the specific treatment of COVID-19 continues to evolve and the most up-to-date guidelines should be reviewed by clinicians involved in management decisions. As noted earlier, in the absence of pediatric-specific clinical trials for the treatment of COVID-19 much of the current guidance involves extrapolation from adult data as well as expert consensus. As detailed later, data from clinical trials in adult patients suggest a benefit to mortality outcomes with the use of dexamethasone, tocilizumab, and janus kinase inhibitors in certain subgroups of patients, potential benefit from the use of remdesivir, as well as emerging evidence for COVID-19 specific monoclonal antibody use. Given the paucity of pediatric data, guidelines recommend specific antiviral therapy should be discussed with the multidisciplinary team including infectious diseases experts and where possible children should be enrolled in available clinical trials [9]. It may be considered more reasonable to implement adult-based recommendations and guidance in older children and adolescents. Some pediatric guidelines suggest on a case-by-case basis specific therapy may be considered appropriate in children with severe or critical COVID-19 or in cases of mild and moderate disease where risk factors for disease progression exist [7–9,59].

Specific treatments for COVID-19 can be broadly categorized into therapies with antimicrobial action, including antiviral therapies; anti-SARS-CoV-2 antibody-based therapies; and those that act mainly via immunomodulatory mechanisms, although considerable overlap may exist for some agents. Broadly speaking, therapies with antiviral action are likely to be most beneficial in the early phases of COVID-19, whereas immunomodulatory agents may be indicated at later stages if aberrant hyperinflammatory responses are present. Specific therapies for which evidence of efficacy for use in patients with COVID-19 exists are discussed later with an overview of the limited evidence available for use in children. A brief discussion of therapies for which there is no clear evidence of benefit is given for some selected agents; however, this list is not exhaustive due to the very large number of investigational agents that have been considered in the therapeutic treatment of COVID-19 patients. For more detail on all investigational and emerging therapies for COVID-19, the reader should refer to regularly updated living evidence summaries, including but not limited to those provided by COVID-NMA, an international initiative working in conjunction with the WHO or by the Australian national COVID-19 taskforce [40,70].

- Agents with antimicrobial mode of action
  - Remdesivir

Box 7. Current use of remdesivir in children is based on a case-by-case assessment in line with national guidelines with input from pediatric infectious diseases experts. This is due to a lack of evidence supporting remdesivir use in children and adolescents with acute COVID-19 leading to variation in national guidelines worldwide. Evidence from adult studies suggests possible benefit in patients requiring supplemental oxygen, but unlikely benefit and possible harm in patients already requiring mechanical ventilation.

Remdesivir is a nucleotide prodrug that inhibits viral RNA-dependent RNA polymerase [71]. Not specifically designed for use against coronaviruses, remdesivir was first utilized clinically in patients with Ebola virus [72]. Early in the pandemic, remdesivir was identified to have direct antiviral activity against coronaviruses in both in vitro and animal model work [71,73,74]. A number of clinical studies on remdesivir use in adult patients with COVID-19 have been published examining its efficacy for both severe and nonsevere hospitalized patients [75–78]. The US NIH-sponsored Adaptive COVID-19 Treatment Trial (ACTT-1) assessed the use of remdesivir versus placebo in 1062 patients admitted to hospital with COVID-19 (laboratory confirmed) and oxygen saturations (O_2 sats) less than 94% on room air [76]. In this study, the use of remdesivir was demonstrated to shorten the median recovery time from a median of 15 days in the placebo arm to 10 days in the remdesivir recipient arm [76]. There was a trend toward reduced mortality in the remdesivir arm; however, this was not significant. Notably, the open-label WHO-sponsored multinational SOLIDARITY trial of hospitalized adults with COVID-19 also found no benefit in reducing 28-day mortality with remdesivir use [78]. More recently, the Open-label, European randomized DisCoVeRy trial of remdesivir versus standard care in hospitalized adults with moderate or severe COVID-19 did not find any significant difference in clinical status at day 15 or in mortality between arms [79]. These trials evaluated the use of a 10-day course of remdesivir; however, an open-label randomized trial of approximately 400 adult patients with COVID-19 requiring oxygen support observed no significant difference between a 10 day versus 5 days course of therapy on improvements in time to clinical recovery [75].

In patients with nonsevere COVID-19, Spinner et al. conducted an open-label randomized trial in 584 adult patients with moderate COVID-19 (defined as any radiographic evidence of pulmonary infiltrates and O_2 sats >94% on room air). Noting a statistically significantly higher odds of better clinical status in the 5-day remdesivir group versus standard care, this work did not observe a significant difference in the 10-day course group and thus the results were considered of uncertain clinical significance [77].

A meta-analyses of the use of remdesivir versus standard care pooling data from four clinical trials did not find a significant reduction in time to symptom resolution (OR 0.82, 95% CI 0.64−1.06) or mortality benefit (OR 0.9, 95% CI 0.72−1.11); however, a small reduction in the need for mechanical ventilation was observed (OR 0.75, 95% CI 0.52−0.98) without any reduction in duration of mechanical ventilation [80]. Based on subgroup analyses of the studies described, a number of national guidelines suggest remdesivir may be beneficial in adults with COVID-19 who require oxygen support but do not require invasive or noninvasive ventilation, however for patients on mechanical ventilation, due to lack of evidence for efficacy and concern for possible harm, remdesivir use is not recommended [15,40,81,82]. However, the certainty of the evidence is low, and overall, a clear large clinical benefit with remdesivir use among hospitalized patients has not been demonstrated.

Pediatric interventional trials for remdesivir use in children with COVID-19 are ongoing but data is not yet available. Limited data from small observational studies and case series suggest that remdesivir is well tolerated in children and neonates [83−85]. Although data on the benefit of remdesivir use in children are lacking, it is licensed in a number of countries worldwide for use in children with acute COVID-19 aged over 12 years and weighing 40 kg or more [8,9,82]. Significant variation exists in national recommendations for its use in the pediatric population [8,9,40]. The US NIH recommends remdesivir in hospitalized children with emergent or increasing need for supplemental oxygen who are aged ≥16 years or ≥12 years with risk factors for severe disease [8]. UK guidance suggests to consider its use only in young people hospitalized with COVID-19 aged ≥12 years weighing 40 kg or more needing low flow supplemental oxygen, but not to use in young people and children in hospital and on high-flow nasal oxygen or higher levels of noninvasive/invasive ventilation support, except as part of a clinical trial [82]. Outside of national licensed indications, guidelines suggest remdesivir use in children with acute COVID-19 should occur within the context of a clinical trial or only be considered on a case-by-case basis with input from pediatric infectious diseases experts.

- Chloroquine or hydroxychloroquine with or without azithromycin

Box 8. Existing data suggest that chloroquine and hydroxychloroquine ± azithromycin are not beneficial in the treatment of adult patients with acute COVID-19 and may cause harm. In the absence of pediatric data, their use is avoided in the treatment of children and adolescents with acute COVID-19.

Chloroquine, an antimalarial drug, demonstrated inhibitory effects against coronaviruses including SARS-CoV-2 in in vitro studies [74,86–90]. Hydroxychloroquine, a metabolite of chloroquine, was also found in vitro to have potency against SARS-CoV-2 [91,92]. Initial results of small non-randomized trials of adult patients with COVID-19 by Gautret et al. showed faster viral clearance in patients receiving hydroxychloroquine versus controls [93,94]. However, the large randomized controlled RECOVERY trial of hospitalized adults in the United Kingdom showed hydroxychloroquine did not decrease 28-day mortality versus standard care and indeed, those on hydroxychloroquine had a longer median of hospital stay and were more likely to require invasive mechanical ventilation or die during hospitalization than those who received standard of care [95]. Several additional large randomized controlled trials, including the WHO-sponsored Solidarity Trial Consortium, also failed to show clinical benefit to use of hydroxychloroquine in hospitalized or nonhospitalized patients adults with COVID-19 whether or not combination with azithromycin was used. A number of studies also highlighted the significant risk of cardiac adverse events with its use, particularly combined with azithromycin [78,96–101]. Furthermore, the UK RECOVERY and PRINCIPLE studies demonstrated respectively that use of azithromycin alone in hospital or community settings was not associated with any clinical benefit for adult patients with COVID-19 in comparison to standard care [102,103].

- Ivermectin

Box 9. Presently, most guidelines recommend against the use of ivermectin outside of the context of a clinical trial given insufficient evidence on the use of ivermectin for the treatment of children and adolescents with acute COVID-19.

Ivermectin, an antiparasitic agent used in the treatment of helminthiasis and scabies, has demonstrated in vitro antiviral activity to a number of viruses, including SARS-CoV-2 [104–106]. Several randomized trials and retrospective reviews, a number of which remain only available on nonpeer-reviewed repositories, have examined the use of ivermectin in adult patients with COVID-19. The US NIH review of the use of ivermectin in the treatment of COVID-19 noted that many studies have significant methodological limitations including small sample size, variation in dosing and schedules used, lack of blinding in trial design, variability in the use of comparator drugs and concomitant medications, poorly described disease severity, and outcome measures not clearly defined [81]. Furthermore, concerns have been raised as to whether some studies were indeed conducted as reported [107]. Due to these limitations, the NIH excluded a number of studies when determining their guidance with respect to ivermectin use. Of the randomized controlled trials with the greatest impact on the NIH panel recommendations, most studies showed no significant or minimal evidence of clinical benefit with regard to symptom resolution; time to negative PCR/decline of viral load; COVID-19-related hospitalization; need for ICU admission, invasive ventilation, or death with ivermectin use [108–113]. Accordingly, the certainty of evidence regarding the use of ivermectin in patients with acute COVID-19 is very low; moreover, none of the reported studies have included pediatric patients. The results of ongoing clinical trials are pending; however, presently, based on available evidence, most guidelines recommend against the use of ivermectin outside of the context of a clinical trial.

- Lopinavir/ritonavir

Box 10. Evidence from adult literature suggests no benefit to the use of lopinavir/ritonavir in the treatment of COVID-19.

Lopinavir/ritonavir is an HIV protease inhibitor, which was used in clinical trials as a therapy for adult patients with SARS-CoV-1 and MERS-CoV [114,115]. This therapy, therefore, gained initial prominence as a potential therapeutic agent in COVID-19. However, two large multicenter randomized controlled trials, the UK-based RECOVERY trial and WHO-sponsored Solidarity trial both found no clinical benefit on mortality outcomes with the use of lopinavir/ritonavir [78,116]. Currently, the US NIH recommends against the use of lopinavir/ritonavir for the treatment of both hospitalized and nonhospitalized patients with COVID-19 [81].

- Anti-SARS-CoV-2 antibody-based therapies
  - Anti-SARS-CoV-2 monoclonal antibodies

Box 11. There is insufficient evidence on the use of anti-SARS-CoV-2 monoclonal antibodies in pediatric patients with acute COVID-19. Many national guidelines recommend exceptional use under specified criteria on a case-by-case basis in consultation with pediatric infectious diseases.

Several anti-SARS-CoV-2 specific recombinant human monoclonal antibodies (mAbs) have been developed for use in patients with COVID-19 and act by binding noncompetitively to nonoverlapping epitopes of the SARS-CoV-2 spike protein receptor-binding domain to block viral entry into host cells. Bamlanivimab plus eteseviumab, casirivimab plus imdevimab (Ronapreve/REGEN-COV), and sotrovimab have US Food and Drug Administration (FDA) Emergency Use Authorization (EUA) as of December 2021 [81]. A number of randomized clinical trials using these anti-SARS-CoV-2 mAbs in adult patients with COVID-19 have been published [117−121]. Ambulatory adult patients with mild-to-moderate COVID-19 who received bamlanivimab plus etesevimab were found to have a lower incidence of COVID-19-related hospitalization and death compared to placebo [117]. Similarly, in nonhospitalized patients with symptomatic COVID-19 within 5 days of symptom onset and at least one risk factor for disease progression, sotrovimab was associated with decreased progression to hospitalization or death compared to placebo [118]. REGEN-CoV has also been shown in an outpatient setting to decrease COVID-19-related hospitalization or death from any cause with a relative risk reduction of approximately 70% [122]. In a hospital setting, preprint data for two large randomized clinical trials have indicated that REGEN-COV is beneficial in reducing all-cause mortality among patients who were seronegative at baseline [120,121].

Guidelines worldwide continue to evolve rapidly due to ongoing results of clinical trials. The emergence of novel SARS-CoV-2 variants also has significant impact on mAbs efficacy. *Alpha* and *Delta* variants demonstrate susceptibility in vitro to all the anti-SARS-CoV-2 mAbs discussed earlier [123,124]. The *Beta* and *Gamma* variants demonstrate reduced in vitro susceptibility to bamlanivimab and etesevimab, as well as casirivimab, but the combination of casirivimab/imdevimab and sotrovimab appear to retain activity against these variants [123,124]. The newly identified variant *Omicron* with numerous mutations in the spike protein is predicted to have significantly reduced susceptibility to bamlanivimab/etesevimab and casirivimab/imdevimab, but sotrovimab appears to retain good efficacy [123,125].

Bamlanivimab and etesevimab were evaluated in pediatric patients in the BLAZE-1 randomized clinical trial; however, this involved only small numbers and no pediatric subjects included required hospitalization due to COVID-19 [123]. Due to limited data on the efficacy and safety of anti-SARS-CoV-2 mAbs in the pediatric population, as well as the lack of clear information on determining high-risk pediatric patients, many national guidelines do not recommend routine pediatric use but may suggest consideration on a case-by-case basis in children fulfilling specific criteria (see Table 6.2).

TABLE 6.2 Summary of US, UK, and Australian pediatric guidelines for use of anti-SARS-CoV-2 mAbs as of the date 1st January 2022.

Guideline	Hospitalized patients	Nonhospitalized patients
US NIH(8)	No data to support the use of anti-SARS-CoV-2 mAbs in hospitalized children for COVID-19.	Insufficient evidence to recommend either for or against the use of mAbs. In consultation with a pediatric infectious diseases specialist, bamlanivimab plus etesevimab or casirivimab plus imdevimab can be considered on a case-by-case basis for children who meet the EUA criteria, but should not be considered routine care.
UK RCPCH(9)	Casirivimab/Imdevimab or Sotrovimab should be considered as per national commissioning policy [126] for use in hospitalized children and young people aged 12 −18 years, weighing at least 40 kg, hospitalized due to symptomatic SARS-CoV-2 infection who have negative SARS-CoV-2 antiSpike antibody. Decisions for therapy should be discussed on a case-by-case basis with a pediatric infectious diseases specialist and other specialists involved in the patient's care.	Children and young people aged at least 12 years, weighing at least 40 kg with onset of symptoms of COVID-19 within the last 5 days and a member of a highest risk group as defined by national commissioning policy [127] should be discussed with regional pediatric infectious diseases service.

TABLE 6.2 Summary of US, UK, and Australian pediatric guidelines for use of anti-SARS-CoV-2 mAbs as of the date 1st January 2022.—cont'd

Guideline	Hospitalized patients	Nonhospitalized patients
Australian Taskforce [40]	Consider using, in exceptional circumstances, casirivimab plus imdevimab in seronegative children and adolescents aged 12 years and over and weighing at least 40 kg with moderate to critical COVID-19 who are at high risk of disease progression. Do not use casirivimab plus imdevimab in seropositive children and adolescents hospitalized with moderate to critical COVID-19	Consider using, in exceptional circumstances, casirivimab plus imdevimab within 7 days of symptom onset in children and adolescents aged 12 years and over and weighing at least 40 kg with mild COVID-19 who are at high risk of deterioration. Consider using, in exceptional circumstances, sotrovimab for the treatment of COVID-19 within 5 days of symptom onset in children and adolescents aged 12 years and over and weighing at least 40 kg who do not require oxygen and who are at high risk of deterioration. Do not use casirivimab plus imdevimab or sotrovimab in children under 12 years old without risk factors for deterioration who have mild or asymptomatic COVID-19 outside of randomized clinical trials.

- Convalescent sera and immunoglobulin therapy

Box 12. There is a lack of evidence regarding the efficacy of convalescent sera and immunoglobulin therapy in children and adolescents with acute COVID-19.

In hospitalized adults with acute COVID-19, the RECOVERY, CONCOR-I, and REMAP-CAP trials are the three largest randomized clinical trials of convalescent plasma use [128–130]. These studies were all stopped early due to futility and did not demonstrate benefit with convalescent plasma use in regards to mortality, or any secondary outcomes including hospitalization time or need for mechanical ventilation. The Convalescent Plasma in Outpatients (C3PO) randomized trial evaluated convalescent plasma in nonhospitalized patients with at least one risk factor for progression to severe COVID-19. This trial was also stopped for futility, with no difference observed for the primary outcome of disease progression in the convalescent arm versus placebo [131]. Data on convalescent plasma in pediatric patients with acute COVID-19 are mainly limited to case reports and a systematic review found insufficient clinical information on its safety and efficacy [132].

In the context of acute COVID-19, retrospective observational studies suggested that intravenous immunoglobulin therapy (IVIG) may decrease mortality in patients with COVID-19 [133,134]. However, a recent multicenter randomized trial, the Intravenous immunoglobulins in patients with COVID-19-associated moderate-to-severe acute respiratory distress syndrome (ICAR) study, found that IVIG did not significantly improve outcomes in adult patients with COVID-19 moderate-to-severe ARDS compared to placebo and in fact showed a tendency toward more adverse events [135]. Evidence for the effectiveness of IVIG in children with acute COVID-19 is lacking.

- Immunomodulatory therapies
  - Systemic corticosteroid therapy

Box 13. Although pediatric data on the use of systemic corticosteroids in the management of acute COVID-19 are lacking, based on adult data, many pediatric guidelines include consideration for dexamethasone therapy use in children and adolescents requiring oxygen support due to severe or critical disease.

A number of large randomized clinical trials in adult patients with acute COVID-19 have demonstrated the benefit of using systemic corticosteroids such as dexamethasone in reducing mortality. The UK RECOVERY collaborative Group trial found a lower incidence of death in dexamethasone recipients compared to standard care among those receiving oxygen or invasive mechanical ventilation [136]. The multicenter CoDEX randomized clinical trial in Brazilian adults with moderate to severe ARDS due to COVID-19 demonstrated a significantly higher number of days free from mechanical ventilation in the dexamethasone treatment group versus standard care [137]. The WHO Rapid Evidence Appraisal for COVID-19 Therapies (REACT) Working Group subsequently published a prospective meta-analysis pooling data from seven randomized clinical trials, including those that examined dexamethasone as well as hydrocortisone or methylprednisolone in adults with acute COVID-19 [138]. This meta-analysis found that all-cause mortality at 28 days was significantly lower in the patients who received corticosteroids compared to usual care or placebo.

Trials in children and adolescents are ongoing and presently the benefits and risks of using systemic corticosteroids for the treatment of acute COVID-19 in the pediatric population remain uncertain. However, based on data from the adult population as detailed above, many guidelines worldwide recommend considering the use of dexamethasone on a case-by-case basis (at a dose of 0.15 mg/kg/day to a maximum of 6 mg daily for up to 10 days) for children with COVID-19 who require supportive oxygen therapy due to severe or critical disease [7–9,40]. These guidelines vary with regard to the exact level of oxygen support criteria and the age group to which consideration applies; however, there is general agreement that corticosteroids would not be routinely indicated in children who do not require any supportive oxygen therapy.

- Interleukin-6 inhibitors

Box 14. Evidence for the efficacy and safety of interleukin-6 inhibitors, such as tocilizumab, in children and adolescents with acute COVID-19 is limited. Many national guidelines for pediatric cases suggest use should ideally occur in the context of a clinical trial or on an exceptional case-by-case basis use may be considered on discussion with clinicians with expertise in immunomodulatory therapy.

Severe COVID-19 has been associated with elevations in a number of cytokines and in particular, increased levels of IL-6 have been shown to correlate with mortality in adult patients with COVID-19 [139,140]. Accordingly, a number of interleukin-6 (IL6) inhibitor agents have been considered with respect to the treatment of COVID-19 including the recombinant humanized anti-IL-6 receptor mAbs, tocilizumab and sarilumab, and the anti-IL-6 mAb, siltuximab. Of these, tocilizumab has undergone the most extensive evaluation in clinical trials. Initial randomized studies such as the BACC Bay placebo-controlled trial [141], the French multicenter CORIMUNO-TOCI-1 trial [142], the EMPACTA placebo-controlled trial [143], and COVACTA multinational placebo-controlled trial [144] as well as the STOP-COVID multicenter cohort study [145] produced conflicting results regarding the efficacy of tocilizumab. However, subsequently, two large randomized trials, by the RECOVERY Collaborative and REMAP-CAP groups combined the use of tocilizumab and corticosteroids in adult patients hospitalized with acute COVID-19 and demonstrated reduced mortality in tocilizumab recipients. Specifically, the RECOVERY trial enrolled a subset of hospitalized adult patients with COVID-19 with hypoxia and evidence of systemic inflammation (C-reactive protein $\geq$75 mg/L) [146]. The REMAP-CAP trial included only hospitalized adult patients who were in ICU and within 24 h of commencing respiratory or cardiovascular organ support and showed the use of either tocilizumab or sarilumab improved survival outcomes [147]. Another large randomized clinical trial, the REMDACTA trial, in hospitalized adults with severe COVID-19 examined the use of tocilizumab plus remdesivir versus remdesivir alone and did not find any difference between these groups for the primary outcome of time to hospital discharge or "ready for discharge" [148].

Based on the results of these studies, a number of adult guidelines recommend the use of tocilizumab or where unavailable, sarilumab, in a subset of patients that meet similar criteria to those patients included in the REMAP-CAP and RECOVERY trials, typically in combination with dexamethasone [40,81,82]. However, as for other agents discussed, data on tocilizumab use in pediatric patients for COVID-19 are limited and no children were enrolled in the trials discussed above. Use of tocilizumab in pediatric cases of acute COVID-19 has been reported [29,149] and some pediatric guidelines include consideration for its use in pediatric patients who require supplemental oxygen and have evidence of systemic inflammation [40]. However, information on the safety including secondary infection rate with the use of tocilizumab for pediatric COVID-19 cases is limited. Accordingly, many guidelines recommend tocilizumab use in children and adolescents with acute COVID-19 should occur within a clinical trial and suggest outside of this context its use should be directed on a case-by-case basis following discussion with a clinician experienced in the use of immunomodulatory therapy in children [8,9].

- Janus kinase inhibitors

Box 15. Janus kinase inhibitors have demonstrated mortality benefits in studies of adult patients with COVID-19; however, there is insufficient evidence for the safety and efficacy of use in pediatric patients with acute COVID-19.

Janus kinase (JAK) inhibitors, such as baracitinib, used in the treatment of rheumatoid arthritis are theorized to have immunomodulatory effects in patients with acute COVID-19 as well as direct antiviral effects through interference with viral entry into host cells [150]. Data from the ACTT-2 demonstrated shorter recovery time on the use of baricitinib plus remdesivir compared to remdesivir alone; however, a major limitation is that corticosteroids were not used as standard care in this study [151]. The COV-BARRIER trial of adult patients with COVID-19 found that the use of baracitinib was associated with mortality benefits compared to placebo when added to standard care, including the use of corticosteroids [152]. However, the safety and efficacy of JAK inhibitors such as baracitinib have not been evaluated in pediatric patients, and current guidelines indicate that there is insufficient evidence to recommend for or against its use outside of clinical trials in hospitalized children with acute COVID-19 [8,40].

- Colchicine

Box 16. Evidence from adult literature suggests no benefit to the use of colchicine in the treatment of acute COVID-19 either in an inpatient or outpatient setting.

Colchicine is an antiinflammatory medication, which has been evaluated in both an inpatient and outpatient setting in adult patients with COVID-19. The COLCORONA and PRINCIPLE trials examined the use of colchicine in nonhospitalized patients and failed to show improvements in primary endpoints of death/hospitalization or time to recovery [153,154]. In adults hospitalized with COVID-19, the large multicenter RECOVERY Trial did not show any significant impact on 28-day mortality, duration of hospital stay, or risk of disease progression with the use of colchicine [155]. Data on the use of colchicine in the treatment of pediatric patients with acute COVID-19 are lacking.

Management of hospitalized neonates with acute COVID-19

Hospitalization due to acute COVID-19 in the neonatal population occurs uncommonly and information on management is mainly derived from case reports and series [156]. A systematic review of COVID-19 in neonates found overall good outcomes and most were asymptomatic or developed only mild symptoms with respiratory support rarely required [157]. Similarly, in a UK-based review of early neonatal SARS-CoV-2 infection, only four infants with neonatal SARS-CoV-2 infection received any respiratory support, and none required ventilation [158]. A very small number of case reports describe the use of remdesivir in critically ill neonates with COVID-19 [83,159]; however, the safety and efficacy of remdesivir in this context are unknown. Likewise, the use of corticosteroids in neonates with acute COVID-19 is described, but, as for remdesivir, data on safety and efficacy are limited [160]. Of note, in one report of severe neonatal COVID-19 complicated by a multidrug-resistant superinfection, a 19-day course of oral steroid therapy was used from day 5 of admission and the authors postulated a potential deleterious effect of steroid-related immunosuppression in contributing to this severe infection [161]. In summary, the focus of management in neonates with acute COVID-19 is supportive care with the use of specific therapeutics in this population only described in very exceptional cases.

Therapeutic management approach to nonhospitalized children with acute COVID-19

As highlighted previously, the vast majority of children with acute COVID-19 will recover with supportive care alone and identification of the small number of children who are likely to progress to severe or critical disease remains challenging. However, on a case-by-case basis following discussion with the multidisciplinary team, including experts in infectious diseases, some children with mild-to-moderate COVID-19 may be identified as being at high risk for progression to severe disease. In these exceptional cases, some experts suggest consideration may be given to the administration of outpatient therapies [7–9,40,162]. Additional factors, such as the age of the child, time from symptom onset, likely variant causing infection, presence of comorbidities, and potential for drug–drug interactions as well as therapy availability and method of administration should be included as part of the consideration of outpatient therapy [162].

Role of specific therapies in nonhospitalized patients with acute COVID-19

Options considered for outpatient therapy include the use of anti-SARS-CoV-2 mAbs, and a discussion of available guidance on their use in a nonhospitalized setting is detailed earlier. Intravenous remdesivir has recently been studied in a randomized, placebo-controlled trial of nonhospitalized adults within 7 days of symptom-onset, noting an 87% lower risk of the primary outcome of hospitalization or death in those receiving 3 days of remdesivir versus placebo [163]. However, its use in this context for children and adolescents has not been studied.

Inhaled corticosteroids have also been identified as a potential COVID-19 therapy for nonhospitalized patients with risk factors for progression. Data on inhaled budesonide use from two open-label randomized clinical trials in adult outpatients with COVID-19 showed budesonide recipients had reduced COVID-19-related urgent care visits and shorter time to first self-reported recovery in the STOIC and PRINCIPLE trials, respectively [164–166]. However, inhaled ciclesonide was studied in two placebo-controlled randomized clinical trials in adult outpatients with COVID-19 and did not achieve the primary efficacy endpoints in either study [167,168]. These studies did not include pediatric cases and therefore conclusions on the efficacy of inhaled corticosteroids for pediatric outpatients with COVID-19 cannot be made at this time. Based on available evidence, UK, US, and Australian guidelines currently differ as to whether or not they recommend the use of budesonide for outpatient treatment of COVID-19 in adult and pediatric cases [40,81,82].

Selective serotonin reuptake inhibitors and in particular fluvoxamine have been postulated to have benefits for use as a therapy in patients with acute COVID-19 via a number of potential mechanisms, such as via binding to the sigma-1 receptor which plays a role in SARS-CoV-2 replication [169]. Two randomized clinical placebo-controlled trials [170,171] have examined the impact of fluvoxamine in high-risk adults with COVID-19 in an outpatient setting. Lenze et al. in a small trial of 152 adult outpatients showed patients treated with fluvoxamine had a lower likelihood of clinical deterioration over 15 days [170]. The larger TOGETHER trial found treatment with fluvoxamine in high-risk outpatients with COVID-19 reduced the need for a composite outcome of retention in a COVID-19 emergency setting or transfer to a tertiary hospital. However, on review of the effect on hospitalizations only, a significant difference was not observed and treatment with fluvoxamine did not demonstrate any beneficial effect on 28-day mortality outcome [171]. Given this low certainty evidence, most adult guidelines currently recommend against fluvoxamine use in the treatment of acute COVID-19 outside of a clinical trial setting. In the absence of any data on its use for the treatment of acute COVID-19 in the pediatric population similar guidance is applied.

Additional novel antiviral agents being studied for use in adult outpatients with acute COVID-19 at high risk of progression to the severe disease include nirmatrelvir-ritonavir (Paxlovid), an oral viral protease inhibitor combination therapy and molnupiravir (Lagevrio), a nucleoside analog, which inhibits SARS-CoV-2 replication. Molnupiravir may affect bone and cartilage growth and therefore is not approved for use in patients younger than 18 years old [172]. Interim results from the EPIC-HR randomized study on the use of Paxlovid in nonhospitalized adults with COVID-19 are currently not available in peer-reviewed publications. Preliminary data releases have indicated that Paxlovid given within 3 days of symptom onset reduced the risk of COVID-19-related hospitalization or death by 89% compared to placebo [173,174]. In absence of peer-reviewed information and pediatric data, it is too early to draw conclusions on the potential role of this novel antiviral drug in nonhospitalized children and adolescents with acute COVID-19.

Additional outpatient management considerations

Supportive care is the mainstay of management for children and adolescents with COVID-19 in the outpatient setting. Current guidelines recommend caregivers should be given appropriate advice and support regarding monitoring of symptoms and signs of clinical deterioration and that isolation to avoid onward transmission should occur according to regional and national guidelines [8].

Case examples: therapy considerations by case severity

TABLE 6.3 **Summary of specific therapy considerations in pediatric acute COVID-19 cases based on current UK, US, Australian, and Italian national guidelines as of the date 1st January 2022. Please note that given the paucity of pediatric data, the use of specific therapy in the context of acute COVID-19 is not considered routine and guidelines recommend decisions regarding specific therapy use should be made on a case-by-case basis on review of most up to date regional/national guidance and with input from infectious diseases experts.**

Disease severity[a]	Therapy considerations[b]	Additional specific therapy information
Mild to moderate/ nonsevere case	• **Supportive care** is mainstay of managementSpecific therapy • **Anti-SARS-CoV-2 mAbs**: In exceptional circumstances in patients with known risk factors for severe disease some guidelines suggest consideration of anti-SARS-CoV-2 mAbs under specified criteria • **Remdesivir:** Some guidelines suggest consideration of remdesivir in exceptional cases with emergent/increasing need for supplemental oxygen under specified age and/or risk criteria.	**Anti-SARS-CoV-2 mAbs** • Approval for specific anti-SARS-CoV-2 mAbs use differs between countries and is evolving rapidly. Current national guidance should be referred to for dosing and administration criteria including; age, time from symptom onset and risk factor criteria. • Available agents for treatment of patients with acute COVID-19 include bamlanivimab/etesevimab, casirivimab/imdevimab and sotrovimab • In vitro susceptibility to the above agents is retained for *Alpha* and *Delta* SARS-CoV-2 variants. *Beta* and *Gamma* variants demonstrate reduced in vitro susceptibility to bamlanivimab/etesevimab. Bamlanivimab/ etesevimab and casirivimab/imdevimab, likely have decreased activity against the newly identified *Omicron* variant however sotrovimab appears to retain activity against this variant. • *Important reported adverse effects*[c]: serious hypersensitivity reactions including anaphylaxis, infusion-related reactions including nausea, chills, dizziness, rash, urticaria, flushing [123,124,126,127]. **Remdesivir** • Dosing: *$\geq$40 kg*: 200 mg IV q24h x1, then 100 mg IV q24h; *$\geq$3.5 to <40 kg*: 5 mg/kg IV q24h x1, then 2.5 mg/kg IV q24h[d]
Severe disease case	• **Supportive care** remains mainstay of management • **Where possible enroll patients in available clinical trials** Specific therapy • **Dexamethasone:** All guidelines recommend consideration of dexamethasone use in hospitalized children who require oxygen therapy support (US guidelines do not routinely recommend for pediatric patients who require only low flow level of oxygen support). • **Remdesivir:** US guidelines suggest consideration of remdesivir in hospitalized children with emergent or increasing need for supplemental oxygen aged $\geq$16 years or $\geq$12 years with risk factors	

	for severe disease. UK guidance suggests to consider use in children hospitalized with COVID-19 aged ≥12 years weighing ≥40 kg requiring low flow supplemental oxygen. • **Anti-SARS-CoV-2 mAbs**: Some guidelines suggest to consider using, in an exceptional case-by-case basis, specific anti-SARS-CoV-2 mAbs in seronegative children and adolescents aged 12 years and over and weighing at least 40 kg hospitalized with symptomatic COVID-19. • **Tocilizumab:** Many national guidelines suggest tocilizumab use in children with acute COVID-19 should occur in a clinical trial and outside of this context should only be considered on an exceptional case-by-case basis. In the largest adult COVID-19 trial of tocilizumab treatment, patients had hypoxia and evidence of systemic inflammation (CRP ≥75 mg/L) and most patients also received systemic corticosteroids as part of standard of care.	• Most guidelines suggest use of a 5-day total course of remdesivir, but up to 10 days were used in some clinical trials and it is unclear which duration provides optimum benefit. • *Important reported adverse effects*[c]*:* Hypersensitivity reactions, including infusion-related and anaphylactic reactions, rash, transaminase elevations [175]. **Dexamethasone** • Dosing: 150 μg/kg orally or IV q24h (max daily dose 6 mg) • Duration: up to 10 days recommended by most guidelines • Note: if dexamethasone is unavailable, guidelines suggests alternate corticosteroids at equivalent dosing, such as prednisone, methylprednisolone, or hydrocortisone, may be considered. **Tocilizumab** • Tocilizumab is administered by IV infusion over 60 min. Dosing may vary according to regional and national guidelines. • *Important reported adverse effects*[c]*:* Serious and sometimes fatal infections have been reported in patients receiving immunosuppressive agents including tocilizumab. Gastrointestinal perforation and hepatic injury have been reported in patients taking tocilizumab. Patients receiving tocilizumab in clinical trials have been reported to have higher rates of neutropenia, thrombocytopenia and elevations of AST or ALT. Hypersensitivity reactions including anaphylaxis have been reported. **Interactions** • Drug–drug interactions should be considered when administering specific therapies in the setting of acute COVID-19. • Available drug interaction checkers can aid with interaction assessments e.g., The University of Liverpool interaction checker, available from: http://www.covid19-druginteractions.org
Critical Disease case	• **Supportive care** remains mainstay of management • **Where possible enroll patients in available clinical trials** Specific therapy • **Dexamethasone:** All guidelines recommend consideration of dexamethasone use in hospitalized children who require critical care level oxygen therapy support. • **Remdesivir:** US guidelines suggest consideration of remdesivir in hospitalized children with emergent or increasing need for supplemental oxygen aged ≥16 years or ≥12 years with risk factors for severe disease. UK guidance suggests to consider use in children hospitalized with COVID-19 aged ≥12 years weighing ≥40 kg requiring low flow supplemental oxygen but not to use in young people and children in hospital on HF nasal oxygen, CPAP, noninvasive MV or invasive MV, except as part of a clinical trial. Of note evidence from adult studies suggests remdesivir is unlikely to be beneficial and may cause harm in patients already requiring mechanical ventilation.	

Continued

TABLE 6.3 Summary of specific therapy considerations in pediatric acute COVID-19 cases based on current UK, US, Australian, and Italian national guidelines as of the date 1st January 2022. Please note that given the paucity of pediatric data, the use of specific therapy in the context of acute COVID-19 is not considered routine and guidelines recommend decisions regarding specific therapy use should be made on a case-by-case basis on review of most up to date regional/national guidance and with input from infectious diseases experts.—cont'd

Disease severity[a]	Therapy considerations[b]	Additional specific therapy information
	• **Anti-SARS-CoV-2 mAbs**: Some guidelines suggest to consider using, in an exceptional case-by-case basis, specific anti-SARS-CoV-2 mAbs in seronegative children and adolescents aged 12 years and over and weighing at least 40 kg hospitalized with symptomatic COVID-19. • **Tocilizumab:** Many national guidelines suggest tocilizumab use in children with acute COVID-19 should occur in a clinical trial and outside of this context should only be considered on an exceptional case-by-case basis. In the largest adult COVID-19 trial of tocilizumab treatment, patients had hypoxia and evidence of systemic inflammation (CRP $\geq$75 mg/L) and most patients also received systemic corticosteroids as part of standard of care.	

Abbreviations: *CPAP*, continuous positive airway pressure; *CRP*, C-reactive protein, *HF*, High-flow; *IV*, intravenous; *mAbs*, monoclonal antibodies; *MV*, mechanical ventilation.
[a] *For details on definitions of disease severity please refer to severity section in the text.*
[b] *For details regarding supportive care and specific therapies please refer to relevant section in the text [7–9,40]: therapies with antiviral action are likely to be most beneficial in early phases of COVID-19, and immunomodulatory agents are more likely to be of benefit in the presence of hyperinflammation.*
[c] *For additional details on drug specific side effects and cautions please refer to relevant summary of product characteristics.*
[d] *Dosing as per FDA remdesivir fact sheet [175].*

Summary

The most effective therapeutic strategy in pediatric COVID-19 remains poorly defined due to a lack of randomized clinical trials involving children. SARS-CoV-2 infection is typically less likely to be severe in children and adolescents compared to adults and thus the vast majority of children will not require specific therapy other than supportive care. In the absence of adequate data, most recommendations in pediatric guidance are based on efficacy and safety data from adult studies. A number of national pediatric guidelines are available, and these continue to be updated regularly with new publications. Based on these guidelines, Table 6.3 provides a broad overview of specific therapy considerations for the management of pediatric cases of acute COVID-19 by disease severity. Improved consensus on the best management approach in pediatric patients with acute COVID-19 is likely to be achieved as further relevant pediatric data emerges and the clinician should remain alert to ongoing updates on management as evidence evolves.

Acknowledgments

The authors would like to acknowledge the expertise and contribution of Dr. Leo Brandao in preparing this chapter.

References

[1] Zimmermann P, Curtis N. Coronavirus infections in children including COVID-19: an overview of the epidemiology, clinical features, diagnosis, treatment and prevention options in children. Pediatr Infect Dis J 2020;39(5):355–68.

[2] Mansourian M, Ghandi Y, Habibi D, Mehrabi S. COVID-19 infection in children: a systematic review and meta-analysis of clinical features and laboratory findings. Arch Pediatr 2021;28(3):242–8. Available from: https://doi.org/10.1016/j.arcped.2020.12.008.

[3] Duarte-Salles T, Vizcaya D, Pistillo A, Casajust P, Sena AG, Hui Lai LY, et al. Thirty-day outcomes of children and adolescents with COVID-19: an international experience. Pediatrics 2021;148(3).

[4] Kitano T, Kitano M, Krueger C, Jamal H, Al Rawahi H, Lee-Krueger R, et al. The differential impact of pediatric COVID-19 between high-income countries and low- and middle-income countries: a systematic review of fatality and ICU admission in children worldwide. PLoS One 2021;16(1 January):1–12.

[5] Irfan O, Muttalib F, Tang K, Jiang L, Lassi ZS, Bhutta Z. Clinical characteristics, treatment and outcomes of paediatric COVID-19: a systematic review and meta-analysis. Arch Dis Child 2021;106(5):440–8.

[6] WHO. Living guidance for clinical management of COVID-19. World Health Organization; 2021. Available from: https://www.who.int/publications/i/item/WHO-2019-nCoV-clinical-2021-1.

[7] Venturini E, Montagnani C, Garazzino S, Donà D, Pierantoni L, Vecchio AL, et al. Treatment of children with COVID-19: update of the Italian society of pediatric infectious diseases position paper. Ital J Pediatr 2021;47(1):10–3.

[8] NIH. COVID-19 treatment guidelines: special considerations in children. National Institutes of Health; 2021. Available from: https://www.covid19treatmentguidelines.nih.gov/special-populations/children/.

[9] RCPCH. COVID-19 - guidance for management of children admitted to hospital. Royal College of Paediatrics and Child Health; 2021. Available from: https://www.rcpch.ac.uk/resources/covid-19-guidance-management-children-admitted-hospital.

[10] Kushner L, Schroeder A, Kim J, Matthew R. "For COVID" or "with COVID": classification of SARS-CoV-2 hospitalizations in children. Hosp Pediatr 2021;11(8).

[11] Dulek DE, Fuhlbrigge RC, Tribble AC, Connelly JA, Loi MM, Chebib H El, et al. Multidisciplinary guidance regarding the use of immunomodulatory therapies for acute COVID-19 in pediatric patients. J Pediatric Infect Dis Soc 2020. https://doi.org/10.1093/jpids/piaa098.

[12] Chiotos K, Hayes M, Kimberlin DW, Jones SB, James SH, Pinninti SG, et al. Multicenter initial guidance on use of antivirals for children with COVID-19/SARS-CoV-2. J Pediatric Infect Dis Soc 2020;9(6):701−15. Available from: https://doi.org/10.1093/jpids/piaa045.

[13] Qiu H, Wu J, Hong L, Luo Y, Song Q, Chen D. Clinical and epidemiological features of 36 children with coronavirus disease 2019 (COVID-19) in Zhejiang, China: an observational cohort study. Lancet Infect Dis 2020;20(6):689−96. Available from: https://doi.org/10.1016/S1473-3099(20)30198-5.

[14] Buonsenso D, Parri N, De Rose C, Valentini P. Toward a clinically based classification of disease severity for paediatric COVID-19. Lancet Infect Dis 2021;21(1):22.

[15] IDSA guidelines on the treatment and management of patients with COVID-19. America, Infectious Diseases Society of; 2021. Available from: https://www.idsociety.org/practice-guideline/covid-19-guideline-treatment-and-management/.

[16] Dong Y, Mo X, Hu Y, Qi X, Jiang F, Jiang Z, et al. Epidemiological characteristics of 2143 pediatric patients with 2019 coronavirus disease in China. Pediatrics. 2020. Available from: http://www.ncbi.nlm.nih.gov/pubmed/32179660.

[17] Nayak D, Roth TL, Mcgavern DB. Pediatric ARDS: consensus recommendations from the pediatric acute lung injury consensus conference. Pediatr Crit Care Med 2015;16(5):428−39.

[18] Lee Y-I, Peng C-C, Chiu N-C, Huang DT-N, Huang F-Y, Chi H. Risk factors associated with death in patients with severe respiratory syncytial virus infection. J Microbiol Immunol Infect October 2014;31 [cited 2014 Dec 9]; 6−11. Available from: http://www.ncbi.nlm.nih.gov/pubmed/25442868.

[19] Fauroux B, Gouyon J-B, Roze J-C, Guillermet-Fromentin C, Glorieux I, Adamon L, et al. Respiratory morbidity of preterm infants of less than 33 weeks gestation without bronchopulmonary dysplasia: a 12-month follow-up of the CASTOR study cohort. Epidemiol Infect July 2014;142(7):1362−74 [cited 2015 Jan 5]; Available from: http://www.ncbi.nlm.nih.gov/pubmed/24029023.

[20] Chan PWK, Lok FYL, Khatijah SB. Risk factors for hypoxemia and respiratory failure in respiratory syncytial virus bronchiolitis. Southeast Asian J Trop Med Publ Health December 2002;33(4). 806−10. [cited 2015 Jan 5]; Available from: http://www.ncbi.nlm.nih.gov/pubmed/12757230.

[21] Grimwood K, Cohet C, Rich FJ, Cheng S, Wood C, Redshaw N, et al. Risk factors for respiratory syncytial virus bronchiolitis hospital admission in New Zealand. Epidemiol Infect October 2008;136(10):1333−41 [cited 2014 Dec 20]; Available from: http://www.pubmedcentral.nih.gov/articlerender.fcgi?artid=2870733&tool=pmcentrez&rendertype=abstract.

[22] Kristensen K, Stensballe LG, Bjerre J, Roth D, Fisker N, Kongstad T, et al. Risk factors for respiratory syncytial virus hospitalisation in children with heart disease. Arch Dis Child

October 2009;94(10):785−9 [cited 2015 Jan 6]; Available from: http://www.ncbi.nlm.nih.gov/pubmed/19541682.

[23] van Beek D, Paes B, Bont L. Increased risk of RSV infection in children with Down's syndrome: clinical implementation of prophylaxis in the European Union. Clin Dev Immunol January 2013;2013:801581. Available from: http://www.pubmedcentral.nih.gov/articlerender.fcgi?artid=3708408&tool=pmcentrez&rendertype=abstract.

[24] Zachariah P, Ruttenber M, Simões EAF. Down syndrome and hospitalizations due to respiratory syncytial virus: a population-based study. J Pediatr May 2012;160(5):827−831.e1 [cited 2015 Jan 6]; Available from: http://www.ncbi.nlm.nih.gov/pubmed/22177993.

[25] Principi N, Esposito S. Severe influenza in children: incidence and risk factors. Expert Rev Anti Infect Ther October 2, 2016;14(10):961−8. Available from: https://doi.org/10.1080/14787210.2016.1227701.

[26] Update: influenza-associated deaths reported among children aged <18 years—United States, 2003−04 influenza season. MMWR Morb Mortal Wkly Rep January 2004;52(53):1286−8.

[27] Garcia MN, Philpott DC, Murray KO, Ontiveros A, Revell PA, Chandromohan L, et al. Clinical predictors of disease severity during the 2009−2010 A(HIN1) influenza virus pandemic in a paediatric population. Epidemiol Infect 2015;143(14):2939−49. Available from: https://www.cambridge.org/core/article/clinical-predictors-of-disease-severity-during-the-20092010-ahin1-influenza-virus-pandemic-in-a-paediatric-population/DB28768F88F9D422C736993EBD1AD021.

[28] Piroth L, Cottenet J, Mariet AS, Bonniaud P, Blot M, Tubert-Bitter P, et al. Comparison of the characteristics, morbidity, and mortality of COVID-19 and seasonal influenza: a nationwide, population-based retrospective cohort study. Lancet Respir Med 2021;2600(20):1−9. Available from: https://doi.org/10.1016/S2213-2600(20)30527-0.

[29] Shekerdemian LS, Mahmood NR, Wolfe KK, Riggs BJ, Ross CE, McKiernan CA, et al. Characteristics and outcomes of children with coronavirus disease 2019 (COVID-19) infection admitted to US and Canadian pediatric intensive care units. JAMA Pediatr September 1, 2020;174(9):868−73. Available from: https://doi.org/10.1001/jamapediatrics.2020.1948.

[30] Drouin O, Hepburn CM, Farrar DS, Baerg K, Chan K, Cyr C, et al. Characteristics of children admitted to hospital with acute sars-cov-2 infection in Canada in 2020. CMAJ (Can Med Assoc J) 2021;193(38):E1483−93.

[31] Swann OV, Holden KA, Turtle L, Pollock L, Fairfield CJ, Drake TM, et al. Clinical characteristics of children and young people admitted to hospital with covid-19 in United Kingdom: prospective multicentre observational cohort study. BMJ 2020:370.

[32] Wolf J, Abzug MJ, Wattier RL, Sue PK, Vora SB, Zachariah P, et al. Initial guidance on use of monoclonal antibody therapy for treatment of coronavirus disease 2019 in children and adolescents. J Pediatric Infect Dis Soc 2021;10(5):629−34.

[33] Bixler D, Miller AD, Mattison CP, Taylor B, Komatsu K. SARS-CoV-2—associated deaths among persons aged <21 Years—United States , February 12−July 31 , 2020. Morb Mortal Wkly Rep 2020;69(37):1324−9.

[34] Ogimi C, Englund JA, Bradford MC, Qin X, Boeckh M, Waghmare A. Characteristics and outcomes of coronavirus infection in children: the role of viral factors and an immunocompromised state. J Pediatric Infect Dis Soc 2019;8(1):21−8.

[35] Nicastro E, Verdoni L, Bettini LR, Zuin G, Balduzzi A, Montini G, et al. COVID-19 in immunosuppressed children. Front Pediatr 2021;9:225. Available from: https://www.frontiersin.org/article/10.3389/fped.2021.629240.

[36] L'Huillier AG, Danziger-Isakov L, Chaudhuri A, Green M, Michaels MG, M Posfay-Barbe K, et al. SARS-CoV-2 and pediatric solid organ transplantation: current knowns and unknowns. Pediatr Transplant August 2021;25(5). e13986−e13986. Available from: https://pubmed.ncbi.nlm.nih.gov/33689201.

[37] El-Sharkawi D, Iyengar S. Haematological cancers and the risk of severe COVID-19: exploration and critical evaluation of the evidence to date. Br J Haematol August 1, 2020;190(3):336−45. Available from: https://doi.org/10.1111/bjh.16956.

[38] Feld LG, Neuspiel DR, Foster BA, Leu MG, Garber MD, Austin K, et al. Clinical practice guideline: maintenance intravenous fluids in children. Pediatrics 2018;142(6).

[39] Intravenous fluid therapy in children and young people in hospital. National Institute for Health and Care Excellence; 2020. Available from: https://www.nice.org.uk/guidance/ng29.

[40] Australian National COVID-19 Clinical Evidence Taskforce. Australian guidelines for the clinical care of people with COVID-19. 2021. Available from: https://app.magicapp.org/#/guideline/L4Q5An.

[41] Alhazzani W, Møller MH, Arabi YM, Loeb M, Gong MN, Fan E, et al. Surviving sepsis campaign : guidelines on the management of critically ill adults with coronavirus disease 2019 (COVID-19). Crit Care Med 2020;46(5):854−87.

[42] Gorman E, Connolly B, Couper K, Perkins GD, McAuley DF. Non-invasive respiratory support strategies in COVID-19. Lancet Respir Med 2021;9(6):553−6.

[43] Weatherald J, Norrie J, Parhar KKS. Awake prone positioning in COVID-19: is tummy time ready for prime time? Lancet Respir Med 2021;9(12):1347−9.

[44] World Health Organisation. Coronavirus disease (COVID-19) advice for the public: When and how to use masks. Available from: https://www.who.int/emergencies/diseases/novel-coronavirus-2019/advice-for-public/when-and-how-to-use-masks.

[45] Ehrmann S, Li J, Ibarra-Estrada M, Perez Y, Pavlov I, McNicholas B, et al. Awake prone positioning for COVID-19 acute hypoxaemic respiratory failure: a randomised, controlled, multinational, open-label meta-trial. Lancet Respir Med 2021;9(12):1387−95.

[46] Leroue M, Maddux A, Mourani P. Prone positioning in children with respiratory failure because of coronavirus disease 2019. Curr Opin Pediatr 2021;33(3):319−24.

[47] Cheifetz IM. Pediatric ARDS. Respir Care 2017;62(6):718−31.

[48] Day M. Covid-19 : ibuprofen should not be used for managing symptoms , say doctors and scientists. Br Med J 2020;1086(March):2020. Available from: https://doi.org/10.1136/bmj.m1086.

[49] Fang L, Karakiulakis G, Roth M. Are patients with hypertension and diabetes mellitus at increased risk for COVID-19 infection? Lancet Respir 2020;2600(20):30116. Available from: https://doi.org/10.1016/S2213-2600(20)30116-8.

[50] Prada L, D Santos C, Baião RA, Costa J, Ferreira JJ, Caldeira D. Risk of SARS-CoV-2 infection and COVID-19 severity associated with exposure to nonsteroidal anti-inflammatory drugs: systematic review and meta-analysis. J Clin Pharmacol December 2021;61(12):1521−33. Available from: https://pubmed.ncbi.nlm.nih.gov/34352112.

[51] Becker RC. COVID-19 update: covid-19-associated coagulopathy. J Thromb Thrombolysis 2020;50(1):54−67. Available from: https://doi.org/10.1007/s11239-020-02134-3.

[52] Cuker A, Tseng EK, Nieuwlaat R, Angchaisuksiri P, Blair C, Dane K, et al. American Society of Hematology living guidelines on the use of anticoagulation for

thromboprophylaxis in patients with COVID-19: may 2021 update on the use of intermediate-intensity anticoagulation in critically ill patients. Blood Adv October 14, 2021;5(20):3951–9. Available from: https://doi.org/10.1182/bloodadvances.2021005493.

[53] Whitworth HB, Sartain SE, Kumar R, Armstrong K, Ballester L, Betensky M, et al. Rate of thrombosis in children and adolescents hospitalized with COVID-19 or MIS-C. Blood 2021;138(2):190–8.

[54] Aguilera-Alonso D, Murias S, Garde AM-A, Soriano-Arandes A, Pareja M, Otheo E, et al. Prevalence of thrombotic complications in children with SARS-CoV-2. Arch Dis Child 2021;106:1129–32.

[55] British Thoracic Society. BTS guidance on venous thromboembolic disease in patients with COVID-19. 2021.

[56] NIH. COVID-19 Treatment Guidelines. Antithrombotic therapy in patients with COVID-19. National Institutes of Health; 2021. Available from: https://www.covid19treatmentguidelines.nih.gov/therapies/antithrombotic-therapy/.

[57] Goldenberg NA, Sochet A, Albisetti M, Biss T, Bonduel M, Jaffray J, et al. Consensus-based clinical recommendations and research priorities for anticoagulant thromboprophylaxis in children hospitalized for COVID-19-related illness. J Thromb Haemostasis November 2020;18(11):3099–105.

[58] Del Borrello G, Giraudo I, Bondone C, Denina M, Garazzino S, Linari C, et al. SARS-COV-2-associated coagulopathy and thromboembolism prophylaxis in children: a single-center observational study. J Thromb Haemostasis February 2021;19(2):522–30.

[59] Beck C, Chauvin-Kimoff L, Krmpotic K, Lirette M, Price V, Thakor S, et al. The acute management of COVID-19 in paediatrics (spring 2021 update). Can Paediatr Soc 2021. Available from: https://cps.ca/documents/position/the-acute-management-of-paediatric-coronavirus-disease-2019covid-19.

[60] Avila L, Amiri N, Pullenayegum E, Sealey VA, De R, Williams S, et al. Diagnostic value of D-dimers for limb deep vein thrombosis in children: a prospective study. Am J Hematol August 1, 2021;96(8):954–60. Available from: https://doi.org/10.1002/ajh.26212.

[61] Monagle P, Cuello CA, Augustine C, Bonduel M, Brandão LR, Capman T, et al. American society of hematology 2018 guidelines for management of venous thromboembolism: treatment of pediatric venous thromboembolism. Blood Adv November 2018;2(22):3292–316.

[62] Langford BJ, So M, Raybardhan S, Leung V, Westwood D, MacFadden DR, et al. Bacterial co-infection and secondary infection in patients with COVID-19: a living rapid review and meta-analysis. Clin Microbiol Infect 2020;26(12):1622–9.

[63] Rawson TM, Moore LSP, Zhu N, Ranganathan N, Skolimowska K, Gilchrist M, et al. Bacterial and fungal coinfection in individuals with coronavirus: a rapid review to support COVID-19 antimicrobial prescribing. Clin Infect Dis November 1, 2020;71(9):2459–68. Available from: https://doi.org/10.1093/cid/ciaa530.

[64] Lansbury L, Lim B, Baskaran V, Lim WS. Co-infections in people with COVID-19: a systematic review and meta-analysis. J Infect August 2020;81(2):266–75. Available from: https://pubmed.ncbi.nlm.nih.gov/32473235.

[65] Raychaudhuri D, Sarkar M, Roy A, Roy D, Datta K, Sengupta T, et al. COVID-19 and co-infection in children: the Indian perspectives. J Trop Pediatr August 27, 2021;67(4):fmab073. Available from: https://pubmed.ncbi.nlm.nih.gov/34478546.

[66] Koehler P, Bassetti M, Chakrabarti A, Chen SCA, Colombo AL, Hoenigl M, et al. Defining and managing COVID-19-associated pulmonary aspergillosis : the 2020 ECMM/ISHAM consensus criteria for research and clinical guidance. Lancet 2021;21:e149–162.

[67] Gangneux J, Dannaoui E, Fekkar A, Luyt C, Botterel F, Prost N De, et al. Articles fungal infections in mechanically ventilated patients with COVID-19 during the first wave: the French multicentre MYCOVID study. Lancet Respir Med 2021;2600(21):1−11.

[68] Diwakar J, Samaddar A, Konar SK, Bhat MD, Manuel E, Hb V, et al. First report of COVID-19-associated rhino-orbito-cerebral mucormycosis in pediatric patients with type 1 diabetes mellitus. J Mycol Med December 2021;31(4):101203. Available from: https://pubmed.ncbi.nlm.nih.gov/34517273.

[69] Nagakumar P, Davies B, Gupta A. Acute asthma management considerations in children and adolescents during the COVID-19 pandemic. Arch Dis Child July 1, 2021;106(7). e31 LP-e31. Available from: http://adc.bmj.com/content/106/7/e31.abstract.

[70] COVID-NMA. The COVID-NMA initiative A living mapping and living systematic review of Covid-19 trials. 2021. Available from: https://covid-nma.com/.

[71] Sheahan TP, Sims AC, Graham RL, Menachery VD, Gralinski LE, Case JB, et al. Broad-spectrum antiviral GS-5734 inhibits both epidemic and zoonotic coronaviruses. Sci Transl Med 2017;9(396).

[72] Wang K, Zhou F, Dingyu Z, Zhao J, Du R, Hu Y, et al. Evaluation of the efficacy and safety of intravenous remdesivir in adult patients with severe pneumonia caused by COVID-19 virus infection: study protocol for a phase 3 randomized, double-blind, placebo-controlled, multicentre trial. BMC trials. 2020. Available from: https://www.researchsquare.com/article/rs-14618/v1.

[73] de Wit E, Feldmann F, Cronin J, Jordan R, Okumura A, Thomas T, et al. Prophylactic and therapeutic remdesivir (GS-5734) treatment in the rhesus macaque model of MERS-CoV infection. Proc Natl Acad Sci USA 2020;117(12):6771−6.

[74] Wang M, Cao R, Zhang L, Yang X, Liu J, Xu M, et al. Remdesivir and chloroquine effectively inhibit the recently emerged novel coronavirus (2019-nCoV) in vitro. Cell Res 2020;30(3):269−71.

[75] Goldman JD, Lye DCB, Hui DS, Marks KM, Bruno R, Montejano R, et al. Remdesivir for 5 or 10 days in patients with severe covid-19. N Engl J Med May 27, 2020; 383(19):1827−37. Available from: https://doi.org/10.1056/NEJMoa2015301.

[76] Beigel JH, Tomashek KM, Dodd LE, Mehta AK, Zingman BS, Kalil AC, et al. Remdesivir for the treatment of covid-19—final report. N Engl J Med 2020;383(19):1813−26.

[77] Spinner CD, Gottlieb RL, Criner GJ, Arribas López JR, Cattelan AM, Soriano Viladomiu A, et al. Effect of remdesivir vs standard care on clinical status at 11 days in patients with moderate COVID-19: a randomized clinical trial. JAMA September 15, 2020;324(11):1048−57. Available from: https://doi.org/10.1001/jama.2020.16349.

[78] Pan H, Peto R, Henao-Restrepo A-M, Preziosi M-P, Sathiyamoorty V. Repurposed antiviral drugs for covid-19—interim WHO solidarity trial results. N Engl J Med December 2, 2021;384(6):497−511. Available from: https://doi.org/10.1056/NEJMoa2023184.

[79] Ader F, Bouscambert-duchamp M, Hites M, Peiffer-smadja N, Poissy J, Belhadi D, et al. Remdesivir plus standard of care versus standard of care alone for the treatment of patients admitted to hospital with COVID-19 (DisCoVeRy): a phase 3, randomised, controlled, open-label trial. Lancet Infect Dis 2021:1−13.

[80] Siemieniuk RAC, Bartoszko JJ, Ge L, Zeraatkar D, Izcovich A, Kum E, et al. Drug treatments for covid-19: living systematic review and network meta-analysis. BMJ July 30, 2020;370:m2980. Available from: http://www.bmj.com/content/370/bmj.m2980.abstract.

[81] National Institutes of Health. COVID-19 treatment guidelines. 2021. Available from: https://www.covid19treatmentguidelines.nih.gov/.

[82] National Institute for Health and Care Excellence. COVID-19 rapid guideline: managing COVID-19. 2021. Available from: https://app.magicapp.org/#/guideline/L4Qb5n/section/ERYAXn.

[83] Saikia B, Tang J, Robinson S, Nichani S, Lawman K-B, Katre M, et al. Neonates with SARS-CoV-2 infection and pulmonary disease safely treated with remdesivir. Pediatr Infect Dis J 2021;40(5). Available from: https://journals.lww.com/pidj/Fulltext/2021/05000/Neonates_With_SARS_CoV_2_Infection_and_Pulmonary.27.aspx.

[84] Méndez-Echevarría A, Pérez-Martínez A, Gonzalez del Valle L, Ara MF, Melendo S, Ruiz de Valbuena M, et al. Compassionate use of remdesivir in children with COVID-19. Eur J Pediatr 2021;180(4):1317−22. Available from: https://doi.org/10.1007/s00431-020-03876-1.

[85] Goldman DL, Aldrich ML, Hagmann SH, Bamford A, Camacho-Gonzalez A, Lapadula G, et al. Compassionate use of remdesivir in children with severe COVID-19. Pediatrics 2021;147(5). e2020047803.

[86] Vincent MJ, Bergeron E, Benjannet S, Erickson BR, Rollin PE, Ksiazek TG, et al. Chloroquine is a potent inhibitor of SARS coronavirus infection and spread. Virol J 2005;2:1−10.

[87] Colson P, Rolain J-M, Lagier J-C, Brouqui P, Raoult D. Chloroquine and hydroxychloroquine as available weapons to fight COVID-19. Int J Antimicrob Agents 2020; 55(4):105932. Available from: http://www.ncbi.nlm.nih.gov/pubmed/32145363.

[88] Keyaerts E, Vijgen L, Maes P, Neyts J, Ranst M Van. In vitro inhibition of severe acute respiratory syndrome coronavirus by chloroquine. Biochem Biophys Res Commun 2004;323(1):264−8.

[89] Kono M, Tatsumi K, Imai AM, Saito K, Kuriyama T, Shirasawa H. Inhibition of human coronavirus 229E infection in human epithelial lung cells (L132) by chloroquine: involvement of p38 MAPK and ERK. Antivir Res 2008;77(2):150−2.

[90] De Wilde AH, Jochmans D, Posthuma CC, Zevenhoven-Dobbe JC, Van Nieuwkoop S, Bestebroer TM, et al. Screening of an FDA-approved compound library identifies four small-molecule inhibitors of Middle East respiratory syndrome coronavirus replication in cell culture. Antimicrob Agents Chemother 2014;58(8):4875−84.

[91] Liu J, Cao R, Xu M, Wang X, Zhang H, Hu H, et al. Hydroxychloroquine, a less toxic derivative of chloroquine, is effective in inhibiting SARS-CoV-2 infection in vitro. Cell Discov 2020;6(16). Available from: https://doi.org/10.1038/s41421-020-0156-0.

[92] Yao X, Ye F, Zhang M, Cui C, Huang B, Niu P, et al. In Vitro antiviral activity and projection of optimized dosing design of hydroxychloroquine for the treatment of severe acute respiratory syndrome coronavirus 2 (SARS-CoV-2). Clin Infect Dis 2020; 71(15):732−9. Available from: https://doi.org/10.1093/cid/ciaa237.

[93] Gautret P, Lagier J-C, Parola P, Hoang VT, Meddeb L, Mailhe M, et al. Hydroxychloroquine and azithromycin as a treatment of COVID-19: results of an open-label non-randomized clinical trial. Int J Antimicrob Agents 2020;56(1):105949. Available from: https://doi.org/10.1016/j.ijantimicag.2020.105949.

[94] Gautret P, Lagier J-C, Parola P, Hoang VT, Meddeb L, Sevestre J, et al. Clinical and microbiological effect of a combination of hydroxychloroquine and azithromycin in 80 COVID-19 patients with at least a six-day follow up: a pilot observational study. Trav Med Infect Dis 2020;34:101663. Available from: https://www.sciencedirect.com/science/article/pii/S1477893920301319?via%3Dihub.

[95] Horby P, Mafham M, Linsell L, Bell JL, Staplin N, Emberson JR, et al. Effect of hydroxychloroquine in hospitalized patients with covid-19. N Engl J Med October 8, 2020;383(21). 2030−40. Available from: https://doi.org/10.1056/NEJMoa2022926.

[96] Self WH, Semler MW, Leither LM, Casey JD, Angus DC, Brower RG, et al. Effect of hydroxychloroquine on clinical status at 14 days in hospitalized patients with COVID-19: a randomized clinical trial. JAMA December 1, 2020;324(21):2165−76. Available from: https://doi.org/10.1001/jama.2020.22240.

[97] Cavalcanti AB, Zampieri FG, Rosa RG, Azevedo LCP, Veiga VC, Avezum A, et al. Hydroxychloroquine with or without azithromycin in mild-to-moderate covid-19. N Engl J Med July 23, 2020;383(21):2041−52. Available from: https://doi.org/10.1056/NEJMoa2019014.

[98] Skipper CP, Pastick KA, Engen NW, Bangdiwala AS, Abassi M, Lofgren SM, et al. Hydroxychloroquine in nonhospitalized adults with early COVID-19. Ann Intern Med July 16, 2020;173(8):623−31. Available from: https://doi.org/10.7326/M20-4207.

[99] Mitjà O, Corbacho-Monné M, Ubals M, Tebé C, Peñafiel J, Tobias A, et al. Hydroxychloroquine for early treatment of adults with mild coronavirus disease 2019: a randomized, controlled trial. Clin Infect Dis December 1, 2021;73(11). e4073−81. Available from: https://doi.org/10.1093/cid/ciaa1009.

[100] Mercuro NJ, Yen CF, Shim DJ, Maher TR, McCoy CM, Zimetbaum PJ, et al. Risk of QT interval prolongation associated with use of hydroxychloroquine with or without concomitant azithromycin among hospitalized patients testing positive for coronavirus disease 2019 (COVID-19). JAMA Cardiol September 1, 2020;5(9):1036−41. Available from: https://doi.org/10.1001/jamacardio.2020.1834.

[101] Nguyen LS, Dolladille C, Drici M-D, Fenioux C, Alexandre J, Mira J-P, et al. Cardiovascular toxicities associated with hydroxychloroquine and azithromycin. Circulation July 21, 2020;142(3):303−5. Available from: https://doi.org/10.1161/CIRCULATIONAHA.120.048238.

[102] Collaborative Group RECOVERY. Azithromycin in patients admitted to hospital with COVID-19 (recovery): a randomised, controlled, open-label, platform trial. Lancet 2021;397:605−12.

[103] PRINCIPLE Trial Collaborative Group. Azithromycin for community treatment of suspected COVID-19 in people at increased risk of an adverse clinical course in the UK (principle): a randomised, controlled, open-label, adaptive platform trial. Lancet 2021;397:1063−74.

[104] Caly L, Druce JD, Catton MG, Jans DA, Wagstaff KM. The FDA-approved Drug Ivermectin inhibits the replication of SARS-CoV-2 in vitro. Antivir Res 2020:104787. Available from: http://www.ncbi.nlm.nih.gov/pubmed/32251768.

[105] Tay MYF, Fraser JE, Chan WKK, Moreland NJ, Rathore AP, Wang C, et al. Nuclear localization of dengue virus (DENV) 1−4 non-structural protein 5; protection against all 4 DENV serotypes by the inhibitor Ivermectin. Antivir Res 2013;99(3):301−6. Available from: https://www.sciencedirect.com/science/article/pii/S0166354213001599.

[106] Barrows NJ, Campos RK, Powell ST, Routh A, Bradrick SS, Garcia-blanco MA, et al. Resource a screen of FDA-approved drugs for inhibitors of Zika virus infection resource A screen of FDA-approved drugs for inhibitors of Zika virus infection. Cell Host Microbe 2016;20:259−70.

[107] Lawrence JM, Meyerowitz-Katz G, Heathers JAJ, Brown NJL, Sheldrick KA. The lesson of ivermectin: meta-analyses based on summary data alone are inherently unreliable. Nat Med 2021;27(11):1853−4. Available from: https://doi.org/10.1038/s41591-021-01535-y.

[108] Vallejos J, Zoni R, Bangher M, Villamandos S, Bobadilla A, Plano F, et al. Ivermectin to prevent hospitalizations in patients with COVID-19 (IVERCOR-COVID19) a randomized, double-blind, placebo-controlled trial. BMC Infect Dis 2021;21(1):635. Available from: https://doi.org/10.1186/s12879-021-06348-5.

[109] López-Medina E, López P, Hurtado IC, Dávalos DM, Ramirez O, Martínez E, et al. Effect of ivermectin on time to resolution of symptoms among adults with mild COVID-19: a randomized clinical trial. JAMA April 2021;325(14):1426–35.

[110] Chowdhury ATMM, Shahbaz M, Karim MR, Islam J, Dan G, He S. A comparative study on ivermectin-doxycycline and hydroxychloroquine-azithromycin therapy on COVID-19 patients. Eurasian J Med Oncol 2021;5(1):63–70.

[111] Mohan A, Tiwari P, Suri TM, Mittal S, Patel A, Jain A, et al. Single-dose oral ivermectin in mild and moderate COVID-19 (RIVET-COV): a single-centre randomized, placebo-controlled trial. J Infect Chemother December 2021;27(12):1743–9.

[112] Galan LEB, Santos NMD, Asato MS, Araújo JV, de Lima Moreira A, Araújo AMM, et al. Phase 2 randomized study on chloroquine, hydroxychloroquine or ivermectin in hospitalized patients with severe manifestations of SARS-CoV-2 infection. Pathog Glob Health June 2021;115(4):235–42.

[113] Ravikirti RR, Pattadar C, Raj R, Agarwal N, Biswas B, et al. Evaluation of ivermectin as a potential treatment for mild to moderate COVID-19: a double-blind randomized placebo controlled trial in Eastern India. J Pharm Pharmaceut Sci 2021;24:343–50.

[114] Chan KS, Lai ST, Chu CM, Tsui E, Tam CY, Wong MML, et al. Treatment of severe acute respiratory syndrome with lopinavir/ritonavir: a multicentre retrospective matched cohort study. Hong Kong Med J 2003;9(6):399–406.

[115] Chu CM, Cheng VCC, Hung IFN, Wong MML, Chan KH, Chan KS, et al. Role of lopinavir/ritonavir in the treatment of SARS: initial virological and clinical findings. Thorax 2004;59(3):252–6.

[116] Collaborative Group RECOVERY. Lopinavir-ritonavir in patients admitted to hospital with COVID-19 (RECOVERY): a randomised, controlled, open-label, platform trial. Lancet (London, England) October 2020;396(10259):1345–52.

[117] Dougan M, Nirula A, Azizad M, Mocherla B, Gottlieb RL, Chen P, et al. Bamlanivimab plus etesevimab in mild or moderate covid-19. N Engl J Med July 14, 2021;385(15): 1382–92. Available from: https://doi.org/10.1056/NEJMoa2102685.

[118] Gupta A, Gonzalez-Rojas Y, Juarez E, Crespo Casal M, Moya J, Falci DR, et al. Early treatment for covid-19 with SARS-CoV-2 neutralizing antibody sotrovimab. N Engl J Med October 27, 2021;385(21). 1941–50. Available from: https://doi.org/10.1056/NEJMoa2107934.

[119] Weinreich DM, Sivapalasingam S, Norton T, Ali S, Gao H, Bhore R, et al. REGN-COV2, a neutralizing antibody cocktail, in outpatients with covid-19. N Engl J Med 2021; 384(3):238–51.

[120] Horby PW, Mafham M, Peto L, Campbell M, Pessoa-Amorim G, Spata E, Staplin N, Emberson JR, Prudon B, Hine P, Brown T. Casirivimab and imdevimab in patients admitted to hospital with COVID-19 (RECOVERY): a randomised, controlled, open-label, platform tri.

[121] Somersan-Karakaya S, Mylonakis E, Menon VP, Wells JC, Ali S, Sivapalasingam S, Sun Y, Bhore R, Mei J, Miller J, Cupelli L. REGEN-COV for treatment of hospitalized patients with covid-19. medRxiv. January 1, 2021.

[122] Weinreich DM, Sivapalasingam S, Norton T, Ali S, Gao H, Bhore R, et al. REGEN-COV antibody combination and outcomes in outpatients with covid-19. N Engl J Med

September 29, 2021;385(23):e81. Available from: https://doi.org/10.1056/NEJMoa2108163.

[123] Food and Drug Administration. Fact sheet for healthcare providers: emergency use authorization (EUA) of bamlanivimab and etesevimab. 2021. p. 2021. Available from: https://www.fda.gov/media/145802/download.

[124] Food and Drug Administration. Fact sheet for healthcare providers: emergency use authorization (EUA) of REGEN-COV (casirivimab and imdevimab). 2021. p. 2021. Available from: https://www.fda.gov/media/145611/download.

[125] Cathcart AL, Havenar-Daughton C, Lempp FA, et al. The dual function monoclonal antibodies VIR-7831 and VIR-7832 demonstrate potent in vitro and in vivo activity against SARS-CoV-2. bioRxiv 2021. Preprint. Available at: https://www.biorxiv.org/content/10.1.

[126] Medicines and healthcare products regulatory agency. Interim clinical commissioning policy: neutralising monoclonal antibodies in the treatment of COVID-19 in hospitalised patients. MHRA 2021. Available from: https://www.cas.mhra.gov.uk/ViewandAcknowledgment/ViewAlert.aspx?AlertID=103187.

[127] Medicines and Healthcare products Regulatory Agency. Commissioning Policy: neutralising monoclonal antibodies or antivirals for non-hospitalised patients with COVID-19. MHRA 2021. Available from: https://www.cas.mhra.gov.uk/ViewandAcknowledgment/ViewAlert.aspx?AlertID=103186.

[128] Bégin P, Callum J, Jamula E, Cook R, Heddle NM, Tinmouth A, et al. Convalescent plasma for hospitalized patients with COVID-19: an open-label, randomized controlled trial. Nat Med November 2021;27(11):2012−24.

[129] Collaborative Group RECOVERY. Convalescent plasma in patients admitted to hospital with COVID-19 (RECOVERY): a randomised controlled, open-label, platform trial. Lancet (London, England) May 2021;397(10289):2049−59.

[130] Estcourt LJ, Turgeon AF, McQuilten ZK, McVerry BJ, Al-Beidh F, Annane D, et al. Effect of convalescent plasma on organ support-free days in critically ill patients with COVID-19: a randomized clinical trial. JAMA November 2021;326(17):1690−702.

[131] Korley FK, Durkalski-Mauldin V, Yeatts SD, Schulman K, Davenport RD, Dumont LJ, et al. Early convalescent plasma for high-risk outpatients with covid-19. N Engl J Med August 18, 2021;385(21):1951−60. Available from: https://doi.org/10.1056/NEJMoa2103784.

[132] Zaffanello M, Piacentini G, Nosetti L, Franchini M. The use of convalescent plasma for pediatric patients with SARS-CoV-2: a systematic literature review. Transfus Apher Sci Off J World Apher Assoc Off J Eur Soc Haemapheresis April 2021;60(2):103043.

[133] Shao Z, Feng Y, Zhong L, Xie Q, Lei M, Liu Z, et al. Clinical efficacy of intravenous immunoglobulin therapy in critical ill patients with COVID-19: a multicenter retrospective cohort study. Clin Transl Immunol January 1, 2020;9(10):e1192. Available from: https://doi.org/10.1002/cti2.1192.

[134] Cao W, Liu X, Hong K, Ma Z, Zhang Y, Lin L, et al. High-dose intravenous immunoglobulin in severe coronavirus disease 2019: a multicenter retrospective study in China. Front Immunol 2021;12:104.

[135] Mazeraud A, Jamme M, Mancusi RL, Latroche C, Megarbane B, Siami S, et al. Intravenous immunoglobulins in patients with COVID-19-associated moderate-to-severe acute respiratory distress syndrome (ICAR): multicentre, double-blind, placebo-controlled, phase 3 trial. Lancet Respir Med 2022;10(2):P158−166.

[136] The RECOVERY Collaborative Group. Dexamethasone in hospitalized patients with covid-19. N Engl J Med July 17, 2020;384(8):693−704. Available from: https://doi.org/10.1056/NEJMoa2021436.

[137] Tomazini BM, Maia IS, Cavalcanti AB, Berwanger O, Rosa RG, Veiga VC, et al. Effect of dexamethasone on days alive and ventilator-free in patients with moderate or severe acute respiratory distress syndrome and COVID-19: the CoDEX randomized clinical trial. JAMA October 6, 2020;324(13):1307−16. Available from: https://doi.org/10.1001/jama.2020.17021.

[138] Group TWHOREA for C-19 T (REACT) W. Association between administration of systemic corticosteroids and mortality among critically ill patients with COVID-19: a meta-analysis. JAMA October 6, 2020;324(13):1330−41. Available from: https://doi.org/10.1001/jama.2020.17023.

[139] Zhou F, Yu T, Du R, Fan G, Liu Y, Liu Z, et al. Clinical course and risk factors for mortality of adult inpatients with COVID-19 in Wuhan, China: a retrospective cohort study. Lancet 2020;6736(20):1−9. Available from: https://doi.org/10.1016/S0140-6736(20)30566-3.

[140] Huang C, Wang Y, Li X, Ren L, Zhao J, Hu Y, et al. Clinical features of patients infected with 2019 novel coronavirus in Wuhan, China. Lancet 2020;395(10223):497−506.

[141] Stone JH, Frigault MJ, Serling-Boyd NJ, Fernandes AD, Harvey L, Foulkes AS, et al. Efficacy of tocilizumab in patients hospitalized with covid-19. N Engl J Med October 21, 2020;383(24):2333−44. Available from: https://doi.org/10.1056/NEJMoa2028836.

[142] Hermine O, Mariette X, Tharaux P-L, Resche-Rigon M, Porcher R, Ravaud P, et al. Effect of tocilizumab vs usual care in adults hospitalized with COVID-19 and moderate or severe pneumonia: a randomized clinical trial. JAMA Intern Med January 1, 2021;181(1):32−40. Available from: https://doi.org/10.1001/jamainternmed.2020.6820.

[143] Salama C, Han J, Yau L, Reiss WG, Kramer B, Neidhart JD, et al. Tocilizumab in patients hospitalized with covid-19 pneumonia. N Engl J Med December 17, 2020;384(1):20−30. Available from: https://doi.org/10.1056/NEJMoa2030340.

[144] Rosas IO, Bräu N, Waters M, Go RC, Hunter BD, Bhagani S, et al. Tocilizumab in hospitalized patients with severe covid-19 pneumonia. N Engl J Med April 2021;384(16):1503−16.

[145] Gupta S, Wang W, Hayek SS, Chan L, Mathews KS, Melamed ML, et al. Association between early treatment with tocilizumab and mortality among critically ill patients with COVID-19. JAMA Intern Med January 1, 2021;181(1):41−51. Available from: https://doi.org/10.1001/jamainternmed.2020.6252.

[146] RECOVERY Collaborative Group. Tocilizumab in patients admitted to hospital with COVID-19 (RECOVERY): a randomised, controlled, open-label, platform trial. Lancet (London, England) May 2021;397(10285):1637−45.

[147] Gordon AC, Mouncey PR, Al-Beidh F, Rowan KM, Nichol AD, Arabi YM, et al. Interleukin-6 receptor antagonists in critically ill patients with covid-19. N Engl J Med April 2021;384(16):1491−502.

[148] Rosas IO, Diaz G, Gottlieb RL, Lobo SM, Robinson P, Hunter BD, et al. Tocilizumab and remdesivir in hospitalized patients with severe COVID-19 pneumonia: a randomized clinical trial. Intensive Care Med 2021;47(11):1258−70. Available from: https://doi.org/10.1007/s00134-021-06507-x.

[149] Derespina KR, Kaushik S, Plichta A, Conway EEJ, Bercow A, Choi J, et al. Clinical manifestations and outcomes of critically ill children and adolescents with coronavirus disease 2019 in New York city. J Pediatr November 2020;226:55–63.e2.

[150] Stebbing J, Phelan A, Griffin I, Tucker C, Oechsle O, Smith D, et al. COVID-19: combining antiviral and anti-inflammatory treatments. Lancet Infect Dis April 2020;20(4):400–2. Available from: https://pubmed.ncbi.nlm.nih.gov/32113509.

[151] Kalil AC, Patterson TF, Mehta AK, Tomashek KM, Wolfe CR, Ghazaryan V, et al. Baricitinib plus remdesivir for hospitalized adults with covid-19. N Engl J Med 2020:1–13. Available from: http://www.ncbi.nlm.nih.gov/pubmed/33306283.

[152] Marconi VC, Ramanan AV, de Bono S, Kartman CE, Krishnan V, Liao R, et al. Efficacy and safety of baricitinib for the treatment of hospitalised adults with COVID-19 (COV-BARRIER): a randomised, double-blind, parallel-group, placebo-controlled phase 3 trial. Lancet Respir Med December 2021;9(12):1407–18.

[153] Tardif J-C, Bouabdallaoui N, L'Allier PL, Gaudet D, Shah B, Pillinger MH, et al. Colchicine for community-treated patients with COVID-19 (COLCORONA): a phase 3, randomised, double-blinded, adaptive, placebo-controlled, multicentre trial. Lancet Respir Med 2021;19(21):1–9. Available from: https://doi.org/10.1016/S2213-2600(21)00222-8.

[154] PRINCIPLE Trial Collaborative Group. Colchicine for COVID-19 in adults in the community (PRINCIPLE): a randomised, controlled, adaptive platform trial. medRxiv 2021. Preprint. Available at: https://www.medrxiv.org/content/10.1101/2.

[155] RECOVERY Collaborative Group. Colchicine in patients admitted to hospital with COVID-19 (RECOVERY): a randomised, controlled, open-label, platform trial. Lancet Respir Med December 2021;9(12):1419–26.

[156] Ryan L, Plötz FB, van den Hoogen A, Latour JM, Degtyareva M, Keuning M, et al. Neonates and COVID-19: state of the art. Pediatr Res 2022;91:432–9. Available from: https://doi.org/10.1038/s41390-021-01875-y.

[157] Trevisanuto D, Cavallin F, Cavicchiolo ME, Borellini M, Calgaro S, Baraldi E. Coronavirus infection in neonates: a systematic review. Arch Dis Child Fetal Neonatal Ed 2021;106(3):F330–5.

[158] Gale C, Quigley MA, Placzek A, Knight M, Ladhani S, Draper ES, et al. Characteristics and outcomes of neonatal SARS-CoV-2 infection in the UK: a prospective national cohort study using active surveillance. Lancet Child Adolesc Heal 2021;5(2):113–21. Available from: https://doi.org/10.1016/S2352-4642(20)30342-4.

[159] Frauenfelder C, Brierley J, Whittaker E, Perucca G, Bamford A. Infant with SARS-CoV-2 infection causing severe lung disease treated with remdesivir. Pediatrics September 1, 2020;146(3):e20201701. Available from: https://doi.org/10.1542/peds.2020-1701.

[160] Trieu C, Poole C, Cron RQ, Hallman M, Rutledge C, Bliton K, et al. Severe neonatal coronavirus disease 2019 presenting as acute respiratory distress syndrome. Pediatr Infect Dis J 2020;39(11). Available from: https://journals.lww.com/pidj/Fulltext/2020/11000/Severe_Neonatal_Coronavirus_Disease_2019.26.aspx.

[161] Marsico C, Capretti MG, Aceti A, Vocale C, Carfagnini F, Serra C, et al. Severe neonatal COVID-19: challenges in management and therapeutic approach. J Med Virol November 23, 2021:1–6. Available from: https://doi.org/10.1002/jmv.27472.

[162] Deville JG, Song E, Ouellette CP. COVID-19: management in children. UpToDate; 2021.

[163] Gottlieb R, Vaca C, Mera P, Webb B, Oguchi P, Ryan P, et al. Early remdesivir to prevent progression to severe covid-19 in outpatients. N Engl J Med 2021:1–11.

[164] PRINCIPLE Collaborative Group. Inhaled budesonide for COVID-19 in people at higher risk of adverse outcomes in the community: interim analyses from the PRINCIPLE trial. medRxiv 2021.

[165] Ramakrishnan S, Nicolau DV, Langford B, Mahdi M, Jeffers H, Mwasuku C, et al. Inhaled budesonide in the treatment of early COVID-19 (STOIC): a phase 2, open-label, randomised controlled trial. Lancet Respir Med 2021;19(21):1−10. Available from: https://doi.org/10.1016/S2213-2600(21)00160-0.

[166] Yu L-M, Bafadhel M, Dorward J, Hayward G, Saville BR, Gbinigie O, et al. Inhaled budesonide for COVID-19 in people at high risk of complications in the community in the UK (PRINCIPLE): a randomised, controlled, open-label, adaptive platform trial. Lancet (London, England) September 2021;398(10303):843−55.

[167] Clemency BM, Varughese R, Gonzalez-Rojas Y, Morse CG, Phipatanakul W, Koster DJ, et al. Efficacy of inhaled ciclesonide for outpatient treatment of adolescents and adults with symptomatic COVID-19: a randomized clinical trial. JAMA Intern Med January 1, 2022;182(1):42−9. Available from: https://doi.org/10.1001/jamainternmed.2021.6759.

[168] Ezer N, Belga S, Daneman N, Chan A, Smith BM, Daniels S-A, et al. Inhaled and intranasal ciclesonide for the treatment of covid-19 in adult outpatients: contain phase II randomised controlled trial. BMJ November 2, 2021;375:e068060. Available from: http://www.bmj.com/content/375/bmj-2021-068060.abstract.

[169] Hashimoto Y, Suzuki T, Hashimoto K. Mechanisms of action of fluvoxamine for COVID-19: a historical review. Mol Psychiatr 2022;27:1898−907. Available from: https://doi.org/10.1038/s41380-021-01432-3.

[170] Lenze EJ, Mattar C, Zorumski CF, Stevens A, Schweiger J, Nicol GE, et al. Fluvoxamine vs placebo and clinical deterioration in outpatients with symptomatic COVID-19: a randomized clinical trial. JAMA December 8, 2020;324(22):2292−300. Available from: https://doi.org/10.1001/jama.2020.22760.

[171] Reis G, Dos Santos Moreira-Silva EA, Silva DCM, Thabane L, Milagres AC, Ferreira TS, et al. Effect of early treatment with fluvoxamine on risk of emergency care and hospitalisation among patients with COVID-19: the together randomised, platform clinical trial. Lancet Global Health January 2022;10(1):e42−51.

[172] FDA News Release. Coronavirus (COVID-19) update: FDA authorizes additional oral antiviral for treatment of COVID-19 in certain adults. US Food and Drug Administration; 2021. Available from: https://www.fda.gov/news-events/press-announcements/coronavirus-covid-19-update-fda-authorizes-additional-oral-antiviral-treatment-covid-19-certain.

[173] Pfizer's novel COVID-19 oral antiviral treatment candidate reduced risk of hospitalization or death by 89% in Interim analysis of phase 2/3 EPIC-HR study. November 05, 2021. Pfizer. 2021. Available from: https://www.pfizer.com/news/press-release/press-release-detail/pfizers-novel-covid-19-oral-antiviral-treatment-candidate.

[174] Mahase E. Covid-19: Pfizer's Paxlovid is 89% effective in patients at risk of serious illness, company reports. BMJ November 8, 2021;375:n2713. Available from: http://www.bmj.com/content/375/bmj.n2713.abstract.

[175] U.S. Food and Drug Administration. Fact sheet for healthcare providers emergency use authorization (EUA) of Veklury (remdesivir) for hospitalized pediatric patients weighing 3.5 kg to less than 40 kg or hospitalized pediatric patients less than 12 years. FDA; 2020. Available from: https://www.fda.gov/media/137566/download.

Chapter 7

COVID-19 vaccinations for children and adolescents

Katrina Nicolopoulos[1,2], Ketaki Sharma[1,2], Lucy Deng[1,2] and Archana Koirala[1,2,3]

[1]*National Centre for Immunisation and Research, Westmead, NSW, Australia;* [2]*The University of Sydney, Sydney, NSW, Australia;* [3]*Nepean Hospital, Kingswood, NSW, Australia*

Vaccination has been one of the most important prevention strategies for many childhood diseases. Although the majority of COVID-19 vaccines are only approved for use in adults aged 18 years and above, an increasing number of vaccines are now also being authorized for use in children.

Vaccination aims to help reduce both direct and indirect harms of COVID-19 in children. COVID-19 is a mild illness for the majority of children [1]. The risk of hospitalization and death is much lower than in the adult population. Some children, however, especially young infants and children with comorbidities, are at risk of hospitalization from severe COVID-19 [2–4]. Protection against COVID-19 aims to minimize children as reservoirs of infection, missed school days through infection, and quarantines. Countries able to procure vaccinations early and complete their adult primary immunization program population have now proceeded to vaccinate children as young as 3 years old.

Vaccine platforms

Globally, there are three types of COVID-19 vaccines used for children: mRNA, inactivated, and protein subunit.

The two mRNA COVID-19 vaccines for children are produced by Pfizer (BNT162b2) and Moderna (mRNA-1273). These vaccines contain mRNA that codes for the full-length SARS-CoV-2 spike protein, enveloped in a lipid nanoparticle. The vaccine is injected into the muscle and the envelope facilitates endocytosis into the host cell, where the mRNA is translated into the spike protein. The spike protein is then released from the cell and acts as a foreign antigen, triggering a host immune response. The mRNA disintegrates rapidly and does not integrate with host DNA. Ultra-cold temperatures are

Clinical Management of Pediatric COVID-19. https://doi.org/10.1016/B978-0-323-95059-6.00002-4

required for the transport of mRNA vaccines, making it difficult to implement in settings without reliable electricity and refrigeration [5]. COVID-19 vaccines using traditional platforms such as inactivated vaccines and protein subunits have been developed and administered to children in countries such as China, India, and Cuba.

A vaccine must go through several phases of clinical trials before it can be licensed.

In phase I clinical trials, typically dozens of participants are recruited. In this phase, the vaccine dose level and safety are tested. In phase II clinical trials, hundreds of participants are recruited. In this phase, the immunogenicity and safety of the vaccine are tested. It is important to ensure that the candidate vaccine stimulates both humoral (reported as geometric mean titer of antibody) and cellular responses against the target antigen. In phase III clinical trials, thousands of participants are recruited. In this phase, the safety and efficacy of the vaccine are tested. The virus must be circulating during the trial to determine if the vaccine is effective to protect against the virus or disease. After collecting data, regulatory bodies assess vaccine safety and effectiveness before the vaccine is licensed. The whole process of vaccine development to licensure takes around 10 years (refer to Fig. 7.1). Once the vaccines were found to be safe and efficacious in adults, trials recruited adolescents, primary school, preschool children, and infants in descending order [6,7].

The majority of COVID-19 vaccines have undergone standardized phased clinical trials to ensure safety, immunogenicity, and efficacy before implementation. Before a vaccine enters clinical trials, it undergoes preclinical assessment, where the target antigen is identified, and the vaccine safety and efficacy are tested in laboratory and animal models. Given the need for rapid vaccine development, clinical trial phases were combined and recruitment into the various trials occurred in parallel (Fig. 7.2).

At the time of publication, there are 10 COVID-19 vaccines approved for use in children in at least one country, utilizing mRNA, viral vector, inactivated, or protein subunit platform technologies. Table 7.1 presents an overview of the available data on these vaccines and on their use in children. Most currently available studies of COVID-19 vaccines in children were conducted before the emergence of the current dominant strains, Delta and Omicron, and therefore there is very limited data on the degree of protection in children against asymptomatic, symptomatic, or severe illness with these newer variants.

mRNA COVID-19 vaccines

The most widely used COVID-19 vaccine in children to date is BNT162b2 (Pfizer/BioNTech), now approved for use in children as young as 5 years old in several countries. This is the only vaccine for which pediatric vaccine effectiveness data are available, though to date only in adolescents. Studies

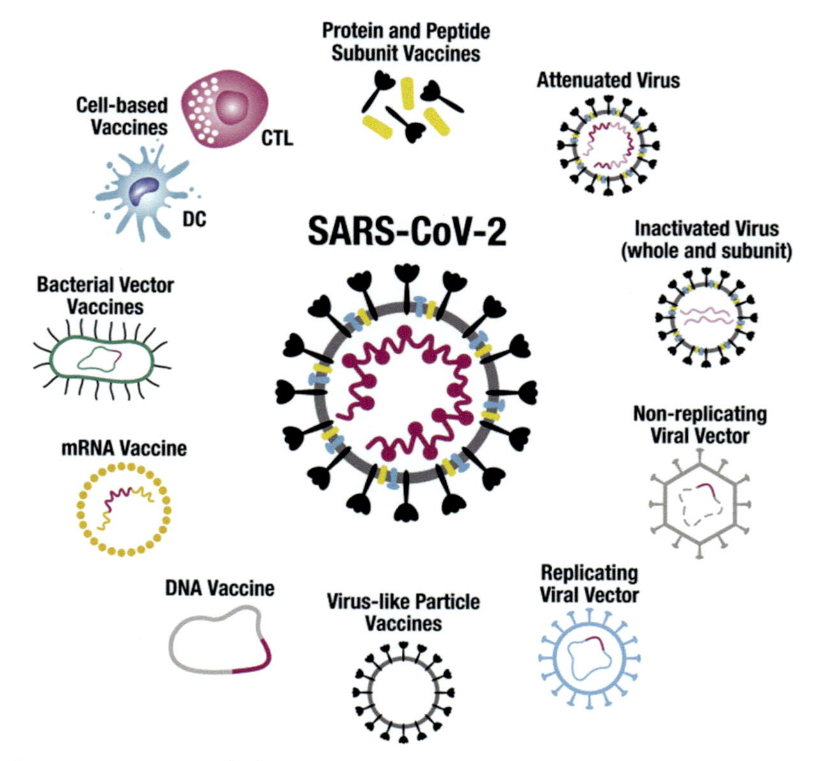

FIGURE 7.1 Illustration of the various COVID-19 vaccine platforms available, and in development [5]. Reproduced with permission: *Flanagan KL, MacIntyre CR, McIntyre PB, Nelson MR. SARS-CoV-2 vaccines: where are we now? J Allergy Clin Immunol Pract. 2021;9(10):3535–43.*

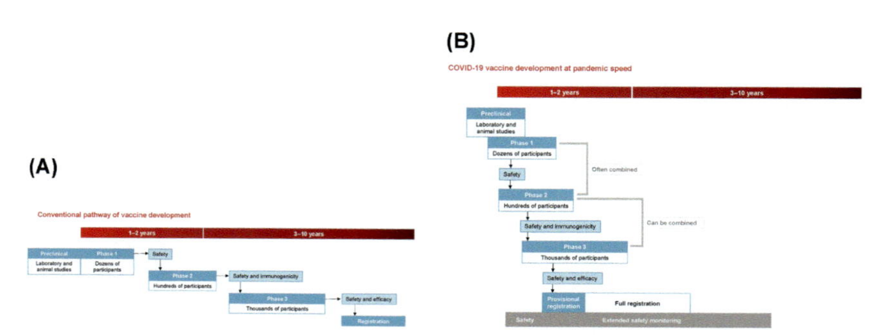

FIGURE 7.2 Pathways of vaccine development: (A) conventional method; (B) COVID-19 vaccine development at pandemic speed. Reproduced, with permission: *Phases of clinical trials. Sydney, Australia: National Center for Immunisation Research and Surveillance; 2020. https://ncirs.org.au/phases-clinical-trials.*

TABLE 7.1 Evidence for the use of COVID-19 vaccines in children (*does not include case reports and case series) as of September 06, 2022.

Vaccine details and countries with approval for use in individuals aged under 18 years	Outcomes	
	Vaccine efficacy (95% CI), effectiveness, immunogenicity	Safety
BNT162b6 [Tozinameran] (Comirnaty) Pfizer/BioNTech Type: mRNA Approvals primary schedule: ***For individuals aged ≥6 months:*** **USA** FDA (June 17, 2022). Coronavirus (COVID-19) Update: FDA aauthorizes Moderna and Pfizer-BioNTech COVID-19 vaccines for children down to 6 months old. Retrieved from: https://www.fda.gov/news-events/press-announcements/coronavirus-covid-19-update-fda-authorizes-moderna-and-pfizer-biontech-covid-19-vaccines-children ***For individuals aged ≥5 years:*** **Australia (41), European Union (42), UK (43), USA (44), Canada (45), New Zealand (46), Mexico (47), Israel (48), UAE (49), Brazil (50), Oman (48), Saudi Arabia, Bahrain, Singapore, Japan, Malaysia, Costa Rica** ***For individuals aged ≥12 years:*** **Switzerland, Norway, South Korea, Philippines, Colombia** ***For individuals aged ≥15 years:*** **Egypt** ***For individuals aged ≥16 years:***	**Vaccine efficacy against laboratory-confirmed COVID-19 ≥7 days after second dose:** Walter et al.: 90.7% (67.7–98.3) *[5–11 years]* (12) Frenck et al.: 100% (75.3–100) *[12–15 years]* (13) Thomas et al.: 91.2% (88.9–93.0) *[≥12 years, until ≥6 months]* (14) Thomas et al.: 91.6% (88.2–94.3) *[≥12 years, at risk with underlying comorbidities]* Thomas et al.: 100% (58.2–100) *[16–17 years, until ≥6 months]* Thomas et al.: 83.7% (74.7–89.9) *[≥12 years, ≥4 months until ≥6 months]* Pfizer Press Release: 73.2% (43.8–87.6) *[3 doses, 6 months to 4 years]* Pfizer Press Release: 75.8% (9.7–94.7) *[3 doses, 6–23 months]* Pfizer Press Release: 71.8% (28.6–89.4) *[3 doses, 2–4 years]* **Vaccine efficacy against severe COVID-19 ≥7 days after second dose:** Thomas et al.: 95.7% (73.9–99.9) *[≥12 years, until ≥6 months]* ***Immunogenicity:*** At 1 month following the second dose, compared to the 16–25-year-old population, noninferiority of immunogenicity analysis was met for the 6–24-month	**Phase 3 safety:** Walter et al. *[5–11 years]*: Favorable safety profile. Reported adverse events generally mild to moderate, and lasting 1–2 days. Systemic events reported more frequently after the second dose, with fatigue (39%) and headache (28%) most frequently reported (12) Frenck et al. *[12–15 years]*: Favorable safety profile, with mainly mild to moderate reactogenicity. Most frequently reported events were injection-site pain (79%–86%), fatigue (60%–66%), and headache (55%–65%). No vaccine-related serious adverse events and few overall severe adverse events (13) FDA Briefing Document *[6 months to 4 years]*: Safety profile was similar to that in other age groups, with more events reported after doses 2 and 3. Lymphadenopathy was reported in 0.2% of vaccine recipients 6–23 months and 0.1% of vaccine recipients 2–4 years (compared to 0% in placebo for both age groups). One case of fever >40.0°C was considered related to the vaccine. There were no febrile convulsions considered related to vaccination. No cases of myocarditis or pericarditis were reported.

Vietnam Approvals booster dose: **For individuals ≥5 years:** **USA, Canada** ***For individuals ≥12 years:*** **Australia, UK, European Union**	old population but not for the 2- to under 5-year-old population [Pfizer Press Release, Dec 17, 2021] (17) Burns et al. *[12–15 years]*: Robust generation of anti-SARS-CoV-2 antibodies following the second dose ($P > .0001$), followed by a significant loss of antibody response after 6 months. At 6 months, most exhibited similar antibody levels to those detected after the first dose, or in the case of anti-Spike antibodies, prevaccine levels of antibodies. This loss of antibody levels highlights a potential vulnerability to breakthrough infections 6 months after their vaccine series (51) **Effectiveness against symptomatic infection:** Reis et al.: 93% (88–97) *[12–18 years]* (9) Glatman-Freedman et al.: 91.5% (88.2–93.9) *[12–15 years]* (10) Powell et al. (8) 93.2% (81.5–97.5) *[12–15 years, Delta, 7–13 days post-2nd dose]* 83.1% (78.2–86.9) *[12–15 years, Omicron, 7–13 days post-2nd dose]* 96.1% (95.2–96.8) *[16–17 years, Delta, 14–34 days post-2nd dose]* 76.1% (73.4–78.6) *[16–17 years, Omicron, 7–13 days post-2nd dose]* 83.7% (72.0–90.5) *[16–17 years, Delta, 70+ days post-2nd dose]* 22.6% (14.5–29.9) *[16–17 years, Omicron, 70+ days post-2nd dose]* Florentino et al.: 62.8% (60.9–64.7) *[12–17 years, Omicron, 14–27 days post-2nd dose]*	**Phase 1 or 2 safety:** With the 3 μg dose, no safety concerns were identified and there was a favorable safety profile in children 6 months to under 5 years old [Pfizer Press Release, Dec 17, 2021] (17) *Phase IV and real-world safety:* Hause et al. *[5–11 years]*: Rates from USA VAERS and v-safe data (53) Hause et al. *[12–17 years]*: Rates from USA VAERS and v-safe data *Myocarditis:* Mevorach et al.: *[12–15 years]* (54) Males: 21 days after first dose: 0.56 cases per 100,000 21 days after second dose: 8.09 cases per 100,000 Females: 21 days after first dose: 0 cases per 100,000 21 days after second dose: 0.69 cases per 100,000 *Booster dose safety:* Hause et al. *[5–11 years]*: Local and systemic reactions were reported with similar frequency after doses 2 and 3. Although local and systemic reactions were similarly reported after receipt of dose 2 and dose 3, some reactions were more frequently reported as moderate or severe after a third than a second dose. Serious adverse events after dose 3 are rare. No reports of myocarditis or death after receipt of dose 3 were received.

Continued

TABLE 7.1 Evidence for the use of COVID-19 vaccines in children (*does not include case reports and case series) as of September 06, 2022.—cont'd

Vaccine details and countries with approval for use in individuals aged under 18 years	Outcomes	
	Vaccine efficacy (95% CI), effectiveness, immunogenicity	**Safety**
	13.9% (10.9–16.9) *[12–17 years, Omicron, ≥98 days post-2nd dose]* Buchan et al.: 51% (38–61) *[12–17 years, Omicron, 7–59 days post-2nd]* 29% (17–38) *[12–17 years, Omicron, ≥180 days post-2nd]* Veneti et al.: 93% (90–95) *[16–17, Delta]* 53% (43–62) *[16–17, Omicron, 7–34 days post-2nd dose]* 23% (3–40) *[16–17, Omicron, ≥63 days post-2nd dose]* Piche-Renaud et al.: 67% (60–72) *[5–11, Omicron]* Cohen-Stavi et al.: 48% (29–63) *[5–11 years, Omicron, 7–21 days after second dose]* Powell et al. 2: 64.5% (63.6–65.4) *[11–17, Omicron, 2 doses]* 62.9% (60.5–65.1) *[11–17, Omicron, 3 doses]* **Effectiveness against any SARS-CoV-2 infection:** Lutrick et al.: 92% (79–97) *[12–17 years]* (52)	

	Cohen-Stavi et al.: 51% (39–61) *[5–11 years, Omicron, 7–21 days after second dose]* **Effectiveness against COVID-19-related hospitalization:** Olson et al.: 94% (90–96) *[12–18 years]* (11) Powell et al.: (8) 83.4% (54.0–94.0) *[12–15 years, Delta, 28+ days post-1st dose]* 76.3% (61.1–85.6) *[16–17 years, Delta, 28+ days post-1st dose]* Price et al.: 92% (80–97) *[12–18 years, Delta, 23–44 weeks post-2nd dose]* 40% (9–60) *[12–18 years, Omicron]* 79% (51–91) *[12–18 years, Omicron, critical COVID-19]* 20% (−25 to 49) *[12–18 years, Omicron, noncritical COVID-19]* 68% (42–82) *[5–11 years, Omicron]* Simmons et al.: 34% (21–66) *[4–17 years, 1 dose]* 56% (17–77) *[4–17 years, 2 doses]* 31% (22–78) *[12–17 years, 1 dose]* 58% (16–80) *[12–17 years, 2 doses]* 34% (−45 to 73) *[4–11 years, 1 dose]* Florentino et al. (hospitalization and death) 75.4% (57.3–85.9) *[12–17 years, Omicron, 14–27 days post-2nd dose]* 84.9% (75.2–90.8) *[12–17 years, Omicron, ≥98 days post-2nd dose]* Buchan et al.: 85% (74–91) *[12–17 years, Omicron, hospitalization, and death]* Piche-Renaud et al.: 94% (56–99) *[5–11, Omicron, hospitalization, and death]* **Effectiveness against COVID-19-related ICU admission:**	

Continued

TABLE 7.1 Evidence for the use of COVID-19 vaccines in children (*does not include case reports and case series) as of September 06, 2022.—cont'd

Vaccine details and countries with approval for use in individuals aged under 18 years	Outcomes	
	Vaccine efficacy (95% CI), effectiveness, immunogenicity	Safety
	Olson et al.: 98% (93–99) *[12–18 years]* (11) **Effectiveness against COVID-19-related hospitalization requiring life-support:** Olson et al.: 98% (92–100) *[12–18 years]* **Use as a booster dose—immunogenicity:** Pfizer Press Release *[5–11 years]*: Data from a subanalysis of 30 sera from a Phase 2/3 clinical trial of children 5 through 11 years old show a 36-fold increase in SARS-CoV-2 Omicron neutralizing titers following a booster dose. Additionally, sera from 140 children receiving a booster dose showed an increase in neutralizing antibodies by sixfold against the SARS-CoV-2 wild-type strain.	
mRNA-1273 [Elasomeran] (Spikevax, TAK-919) Moderna Type: mRNA Approvals: ***For individuals aged ≥6 months:*** **Australia, USA, Canada** ***For individuals aged ≥6 years:*** Australia (55) ***For individuals aged ≥12 years:*** **European Union (56), UK (57), Canada (58), Switzerland (59), Colombia (60)**	**Vaccine efficacy against PCR-confirmed symptomatic infection, ≥14 days after second dose:** Ali et al.: 93.3% (47.9–99.9) *[12–17 year olds]* (38) Creech et al.: 88.0% (70.0–95.8) *[6–11 years olds, after dose 1]* VRBPAC FDA Briefing Document: 50.6% (21.4–68.6) *[6–23 months]* VRBPAC FDA Briefing Document: 36.8% (12.5–54.0) *[2–4 years]* **Vaccine efficacy against PCR-confirmed symptomatic infection, ≥14 days after first dose:** Moderna Press Release, Nov 04, 2021: 100% (no CI) *[6–11 years]*(61)	**Phase 3 safety:** Ali et al. *[12–17 years]*: Reported adverse events lasted a mean of 4 days and were more common after the second dose. Most frequently reported events were injection-site pain (92.4%), headache (70.2%), and fatigue (67.8%). No serious adverse events related to vaccination were reported (38) VRBPAC FDA briefing document: The adverse reaction profile was generally similar to that seen in other age groups. The exception was fever after any dose, which was more frequently reported in those aged 6 months to

	Vaccine efficacy against PCR-confirmed asymptomatic infection, ≥14 days after first dose: Moderna Press Release, Nov 04, 2021: 65% (0.16 –0.85) *[6–11 years]* **Vaccine efficacy against SARS-CoV-2 infection [regardless of symptoms], ≥14 days after first dose:** Moderna Press Release, Nov 04, 2021: 80% (0.62 –0.90) *[6–11 years]* **Immunogenicity:** *[6–11 years]*: Seroresponse rate of 99.3%, with the SARS-CoV-2 neutralizing antibody geometric mean ratio 0.6% (–2.8 to 2.8) greater than phase 3 results in young adults [Moderna Press Release, Oct 25, 2021] [TGA AusPAR, Feb 21, 2022] (62, 63) Bartsch et al. *[6–11 years]*: Compared to fully vaccinated adults, children had significantly reduced SARS-CoV-2 S specific IgM and IgA1 levels after second dose but generated slightly higher IgG1 binding titers. IgG1 titers at both timepoints exceeded titers observed in acute pediatric COVID-19 and MultiInflammatory Syndrome in children (MIS-C). Following second dose, levels of NAb were comparable in 50 μg group, and slightly higher in 100 μg group than levels observed in adults. NAb levels are also higher than in children with acute COVID-19 and MIS-C (64) **Effectiveness against symptomatic infection:** Powell et al. 2: 64.5% (63.6–65.4) *[11–17, Omicron, 2 doses, includes BNT]* 62.9% (60.5–65.1) *[11–17, Omicron, 3 doses, includes BNT]* **Effectiveness against COVID-19-related hospitalization:** Gonzalez et al.: 80.0% (64.3–88.0) *[12–17 year olds, mixed BNT and m1273]*	5 years than in adolescents and young adults. Lymphadenopathy was reported in 1.5% of Moderna recipients and 0.2% of placebo recipients aged 6–23 months, and in 0.9% of Moderna recipients and <0.1% of placebo recipients aged 2–5 years. One vaccine-related febrile convulsion was reported. No cases of myocarditis or pericarditis were reported during the trial. **Phase 1 or 2 safety:** *[6–11 years]:* Generally well tolerated with a safety and tolerability profile mostly consistent with the Phase 3 adolescent and adult trials. The majority of adverse events were mild or moderate in severity, with the most frequently reported solicited adverse events of fatigue, headache, fever, and injection site pain [Moderna Press Release, Oct 25, 2021] (62, 63)

Continued

TABLE 7.1 Evidence for the use of COVID-19 vaccines in children (*does not include case reports and case series) as of September 06, 2022.—cont'd

Vaccine details and countries with approval for use in individuals aged under 18 years	Outcomes	
	Vaccine efficacy (95% CI), effectiveness, immunogenicity	**Safety**
AZD1222/ChAdOx (Vaxzevria, Covishield) University of Oxford/AstraZeneca Type: Viral vector Approvals: **No approvals registered in any country for children <18 years**	**Immunogenicity:** Li et al. *[6–17 years]*: Among participants without prior infection who received ChAdOx, anti-SARS-CoV-2 IgG and pseudoneutralizing antibody titers at day 28 after the second dose were higher in participants aged 12–17 years with a longer interval between doses compared with those aged 12–17 years who received their vaccines 28 days apart. Humoral responses were higher in those aged 6–11 years than in those aged 12–17 years after the second dose. Cellular responses peaked after the first dose of ChAdOx and remained above baseline after the second dose.	**Phase 2 safety:** Li et al. *[6–17 years]*: Of the participants who received ChAdOx, 80% reported at least one solicited local or systemic adverse event up to 7 days following the first dose, 76% following the second dose. No serious adverse events related to ChAdOx administration were reported during the trial. One participant aged 6–11 years receiving ChAdOx reported a grade 4 fever of 40.2°C on day 1 following first vaccination, which resolved within a day. Pain and tenderness were the most common local solicited adverse events.
Gam-COVID-Vac/rAd26-S + rAd5-S (Sputnik V) Gamelaya Institute Type: Viral vector Approvals: ***For individuals aged ≥12 years:*** **Russia [registered as Sputnik M] (65), Kazakhstan (66)**	**Efficacy:** TASS Press Release *[12–17 years]*: Efficacy of Sputnik M against COVID-19 infection is approximately 93% (no CI) [Jan 11, 2022] (67)	Not published or announced
Ad5-nCoV (Convidecia) CanSinoBIO	**Immunogenicity:**	**Phase 1 or 2 safety:**

Type: Viral vector Approvals: ***None reported***	Zhu et al. *[6–17 years]*: Geometric mean titers of antibody were 1037.5 (889.3–1210.5) in those aged *6–17 years* compared to 647.2 (399.3–1049.0) in those aged 18–55 years, 28 days after the second dose. Neutralizing antibody titers were 168.0 (143.3–197.1) in those aged *6–17 years* compared to 76.8 (52.4–112.7) in those aged 18–55 years (24)	Zhu et al. *[6–17 years]*: Within 14 days after each dose at least one adverse event was reported by 69% of participants aged 6–17 years and 70.0% aged 18–55 years. The most reported adverse events were injection-site pain, fever, headache, and fatigue. Grade 3 fever was the only severe adverse event, and was reported in 4% of participants aged 6–17 years, and 10% of participants aged 18–55 years (24)
CoronaVac Sinovac Type: Inactivated Approvals: ***For individuals aged ≥3 years:*** **China (48), Hong Kong (68), Colombia (69)** ***For individuals aged ≥6 years:*** **Brazil (50), Chile (48), El Salvador, Ecuador, Indonesia**	**Immunogenicity:** Han et al. *[3–17 years]*: All participants seroconverted in phase 1. In phase 2, 96.8% (93.1–98.8) of participants in the 1.5 μg group and 100.0% (98.0–100.0) in the 3.0 μg group seroconverted. There were geometric mean titers of 86.4 (73.9–101.0) 1.5 μg group and 142.2 (124.7–162.1) in the 3.0 μg group. There were no detectable antibody responses in the alum-only control groups. The 3.0 μg dose was selected for further study (70) Soto et al. *[3–17 years]*: Four weeks after the second dose, significant increase in the levels of total and neutralizing antibodies, and increase in specific CD4+ T cells activity. NAb titers were lower for Delta and Omicron variants than the D614G variant, but comparable T-cell responses were detected against these variants of concern (71) Wang et al. *[3–17 years]*: By 10 months after dose 2, the GMTs of neutralizing antibody against the prototype SARS-CoV-2 strain were down to 12.7 (95% CI 10.4–15.5) and 20.8 (95% CI 16.5–26.1) in the 1.5 μg	**Phase 1 or 2 safety:** Han et al. *[3–17 years]*: In the combined safety profile of phase 1 and phase 2, any adverse reactions within 28 days after injection occurred in 26% of participants in the 1.5 μg group, 29% in the 3.0 μg group, and 24% in the alum-only group. Most adverse reactions were mild (grade 1) and moderate (grade 2) in severity. There were only <1% grade 3 adverse reactions. Most adverse reactions occurred within 7 days after vaccination and participants recovered within 2 days. The most commonly reported reactions were injection site pain (13%) and fever (5%). In an age subgroup analysis, adverse reactions were more frequently reported in participants aged 12–17 years (35% of participants) followed by 3–5 years (26%) and 6–11 years (18%) (70) Soto et al. *[3–17 years]*: Adverse reactions primarily mild and local, with no severe adverse events reported. The most frequently

Continued

TABLE 7.1 Evidence for the use of COVID-19 vaccines in children (*does not include case reports and case series) as of September 06, 2022.—cont'd

Vaccine details and countries with approval for use in individuals aged under 18 years	Outcomes	
	Vaccine efficacy (95% CI), effectiveness, immunogenicity	**Safety**
	group and 3.0 μg group, respectively. By 12 months after dose 2, the GMTs of neutralizing antibody against the prototype were down to 16.8 (13.3–21.3) and 21.7 (18.1–26.0) in the 1.5 μg group and 3.0 μg group, respectively. **Effectiveness for prevention of COVID-19:** Jara et al.: (23) 74.5% (73.8–75.2) *[6–16 years]* 75.8% (74.7–76.8) *[6–11 years]* **Effectiveness against COVID-19 related hospitalization:** Jara et al.: 91.0% (87.8–93.4) *[6–16 years]* 77.9% (61.5–87.3) *[6–11 years]* 93.8% (87.8–93.4) *[6–16 years,* ICU admission*]*	reported systemic reaction was headache in adolescents, and fever in children aged 3–11 years (first dose: 9% vs. second dose: 7%). Other systemic AEs were reported in less than 10% of vaccinated participants (71)
BBIBP-CorV [Covilo] Sinopharm/Beijing Institute of Biological Products Type: Inactivated Approvals: ***For individuals aged ≥3 years:*** **China (48), Bahrain, Argentina**	**Immunogenicity:** Xia et al. *[3–17 years]*: Neutralizing antibody geometric mean titers ranged from 105.3 to 180.2 in participants aged 3–5 years, 84.1 to 168.6 in the 6–12 years cohort, and 88.0 to 155.7 in the 13–17 years cohort 28 days after the second dose. Neutralizing antibody geometric mean titers ranged from 143.5 to 224.4 in the 3–5 years cohort, 127 to 184.8 in the 6–12 years cohort, and 150.7 to 199 in the 13–17 years cohort, 28 days after the third dose (72)	**Phase 1 or 2 safety:** Xia et al. *[3–17 years]*: Adverse events were mostly mild to moderate in severity. The most frequently reported local adverse event was pain in 4–9.1% of participants in the vaccine group. The most common systematic adverse reaction was fever, reported in 5.2%–12.7% of participants in the vaccine group, and highest in the 3–5 year cohort (72) Tawinprai et al. *[12–17 years]*: The vaccine was not as reactogenic in the 12–17 year

	Tawinprai et al. *[12–17 years]*: The mean concentration of antibody titers in the 12–17 year cohort was 102.9 BAU/mL (91.0–116.4) and 36.9 BAU/mL (30.9–44.0) in the 18–30 year cohort. The geometric mean ratio of the 12–17 year cohort was 2.79 (2.25–3.46, *P*-value; <.0001) compared with the 18–30 year cohort which met the non-inferiority criteria (73) **Effectiveness against COVID-19-related hospitalization:** Gonzalez et al.: 76.4% (62.9–84.5) *[3–11 years]*	cohort compared with the 18–30 year cohort. In both cohorts, the first dose was more reactogenic than the second dose. In the 24 h following the first dose, 18.1% of 12–17 years reported an adverse event compared to 23.7% of 18–30 years. Most reactions were mild to moderate in severity, and resolved within 7 days. Local reactions were reported more frequently than systemic reactions in the 12–17 year cohort, with pain at the injection site the most frequently reported (13.2%). The most common systemic adverse reactions were fatigue (8.7%), myalgia (7.7%), and headache (7.7%). Adverse events were reported approximately 3.4% in the 12–17 year cohort compared to 2% in the adult cohort (73)
WIBP-CorV/COVIV Sinopharm/Wuhan Institute of Biological Products Type: Inactivated Approvals: ***For individuals aged ≥3 years:*** **China (48)**	Not published or announced	Not published or announced
BBV152 (Covaxin) Bharat Biotech Type: Inactivated Approvals: ***For individuals aged ≥12 years:***	**Immunogenicity:** Vadrevu et al. *[2–18 years]*: Seroconversion was between 95% and 98% in all three age sub-groups 4 weeks after the second dose. The geometric mean ratio was 1.76 (1.32–2.33) compared to the adult	**Phase 2–3 safety:** Vadrevu et al. *[2–18 years]*: No serious adverse events, deaths, or withdrawals due to an adverse event reported. The most commonly reported local reaction was

Continued

TABLE 7.1 Evidence for the use of COVID-19 vaccines in children (*does not include case reports and case series) as of September 06, 2022.—cont'd

Vaccine details and countries with approval for use in individuals aged under 18 years	Outcomes	
	Vaccine efficacy (95% CI), effectiveness, immunogenicity	Safety
India (74)	group, and had a lower limit of ≥1, indicating a superior antibody response in children compared to adults that of adults, and meeting noninferiority criteria. Vaccine responses were skewed toward a Th1 response (75)	injection site pain. Systemic adverse events were less frequent, especially after the second dose. The most frequently reported systemic adverse event after dose one was mild-to-moderate fever, reported in 5% of participants aged 12–18 years, 10% of participants 6–12 years, and 13% of 2–6 years. No cases of severe fever was reported, and rates were all 4% or less after the second dose (75) **Phase 4 safety:** Kaur et al. *[15–18 years]*: Adverse events were reported in 36.3% adolescents after first dose and in 37.9% after second dose. Systemic adverse events were reported in 15%–17% adolescents. Injection site pain and fever were the common AEs. Majority were mild to moderate in severity. Most AEs resolved within 1–2 days. No difference in AEFI incidence and patterns was observed between adolescents and adults.
NVX-CoV2373 (Nuvaxovid, Covovax) Novavax Type: Protein subunit Approvals: ***For individuals aged ≥12 years:***	Novavax Press Release *[12–17 years]*: Overall protective efficacy of 79.5% (46.8–92.1) against COVID-19. Efficacy consistent across age groups. Vaccine efficacy against the Delta variant was 82.0% (32.4–95.2). Immune responses were about 2–3-fold higher in adolescents than in adults against all variants	Novavax Press Release *[12–17 years]*: Serious and severe adverse events were low in number and balanced between vaccine and placebo groups, and not considered related to the vaccine. Local and systemic reactogenicity was generally lower than or similar to adults,

Australia, European Union, USA, UK, India, Japan, New Zealand, Switzerland, Thailand, South Korea	studied (76) **Vaccine efficacy against laboratory-confirmed symptomatic COVID-19, ≥7 days after second dose:** Nuvaxovid AusPAR: 79.5% (46.8–92.1) *[12–17 years]* Nuvaxovid AusPAR: 82.0% (32.4–95.2) *[12–17 years, Delta]*	after the first and second dose. The most common adverse reactions observed were injection site tenderness/pain, headache, myalgia, fatigue, and malaise. There was no increase in reactogenicity in younger (12 to <15 years old) adolescents compared to older (15 to <18 years old) adolescents. No safety signal was observed through the placebo-controlled portion of the study (76) **Phase 3 safety:** Nuvaxovid AusPAR: The adverse reaction profile was generally similar to that among adult participants aged 18 years and older. The exception was fever, which was more common among adolescents than adults, with 1% of adolescents reporting fever after the first dose, and 17% after the second dose (2% Grade 3). The frequency of local and systemic reactions was generally similar among participants aged 12–14 years and those aged 15–17 years. There was only one correctly recorded grade 4 event, which was a case of headache following dose 2 in a Novavax COVID-19 vaccine recipient. There were no cases of myocarditis or pericarditis reported up until data cut-off in the phase 3 trial.
FINLAY-FR-02 (Soberana 02) Instituto Finlay de Vacunas Type: Protein subunit Approvals: ***For individuals aged ≥2 years:***	**Immunogenicity:** Puga-Gomez et al. *[3–18 years]*: Two doses elicited a humoral immune response similar to natural infection. The third dose significantly increased the response in all children, similar to that achieved in vaccinated young	**Phase 1 or 2 safety:** Puga-Gomez et al. *[3–18 years]*: Local pain at the injection site was the adverse event reported with the greatest frequency (>10%).

Continued

TABLE 7.1 Evidence for the use of COVID-19 vaccines in children (*does not include case reports and case series) as of September 06, 2022.—cont'd

Vaccine details and countries with approval for use in individuals aged under 18 years	Outcomes	
	Vaccine efficacy (95% CI), effectiveness, immunogenicity	**Safety**
Cuba (77), Venezuela (78)	adults, and higher than the response in convalescent children. The neutralizing titer was evaluated in a small subset of participants: Geometric mean titer (GMT) was 173.8 (95% CI 131.7; 229.5) against alpha, 142 (95% CI 101.3; 198.9) against delta and 24.8 (95% CI 16.8; 36.6) against beta. An estimated efficacy of >90% against symptomatic infection was calculated. CECMED Press Release *[3–18 years]*: Results from the pediatric trial were superior in all immunological variables compared to the adult population aged 19–80 years and similar compared to the subgroup of young adults between 19 and 29 years old (77)	No serious or severe adverse events were reported. CECMED Press Release *[3–18 years]*: The safety profile evidenced was similar between the compared groups (77)
CIGB-66 (Abdala) Center for Genetic Engineering and Biotechnology Type: Protein subunit Approvals: ***For individuals aged ≥2 years:*** **Cuba (79, 80)**	Not published or announced	Not published or announced

SCB-2019 Clover Biopharmaceuticals Type: Protein subunit Approvals: **No approvals registered in any country for children <18 years**	**Immunogenicity:** Clover Biopharmaceuticals Press Release *[12–17 years]*: Elicited approximately twofold higher neutralizing antibody titers in adolescents compared to young adults (aged 18–25 years), a population where the vaccine had previously been demonstrated to be highly protective against COVID-19 [Aug 25, 2022].	**Phase 2 and 3 safety:** Clover Biopharmaceuticals Press Release *[12–17 years]*: Demonstrated a favorable safety and reactogenicity profile. Adverse events were mostly mild and transient, were balanced between vaccine and placebo (saline) groups, and comparable to results observed in the adult population.
BECOV2 (Corbevax) Biological E Limited Type: Protein subunit Approvals: **No approvals registered in any country for children <18 years**	**Immunogenicity:** Thuluva et al. *[5–17 years]*: There were significant increases in antibody and neutralizing antibody titers. Both humoral and cellular immune responses were found to be noninferior to the immune responses induced by CORBEVAX™ vaccine in adult populations.	**Phase 2 and 3 safety:** Thuluva et al. *[5–17 years]*: The safety profile in both pediatric cohorts was comparable to the placebo control group. Majority of reported adverse events (AEs) were mild in severity. No serious AEs, medically attended AEs or AEs of special interest (AESI) were reported during the study period. All the reported AEs resolved.
ZyCoV-D Zydus Cadila Healthcare Type: DNA Approvals: ***For individuals aged $\geq$12 years:*** **India (74)**	Not published or announced	Not published or announced

conducted across various countries and during periods of Delta and Omicron predominance showed that BNT162b2 is highly effective in preventing symptomatic infection and hospitalization in adolescents against older (pre-Omicron) variants of SARS-CoV-2, with efficacy estimates of around 80%−94% [8−10]. Short-term effectiveness against ICU admission is higher at 98% [11]. These studies support the findings of the pivotal clinical trials of this vaccine in children aged 5−18, which demonstrated vaccine efficacy of 90%−100%, including in children with certain medical comorbidities (refer to Table 7.1) [12−14].

There is limited evidence on the duration of protection following COVID-19 vaccination in children and on the need for and timing of booster doses. Protection against symptomatic infection may be short-lived, with studies suggesting waning effectiveness against the Omicron variant by day 70 after the primary course to 22.6% [8]. Early effectiveness study in New York, USA suggests that vaccine efficacy with 2-dose Pfizer vaccine declined rapidly to 51% in 12−17 year olds and 12% in 5−11 year olds by 1 month post-vaccination [15]. A separate study, suggested a vaccine efficacy against infection to be 59% among 12−15 year olds and 31% among 5−11 year olds [16]. However, protection against severe illness is likely to be maintained, based on trends observed in adult populations.

Data on the immunogenicity of BNT162b2 in children aged 6 months to 4 years are anticipated to be published in the coming months, and a press release from the manufacturer indicates that antibody titers induced in the 6−24-month-old population were noninferior to young adults; however, non-inferiority could not be demonstrated in children aged 2-to-5 years old, and the trial will be modified to include a third dose [17].

mRNA-1273 (Moderna) has also demonstrated high efficacy against symptomatic infection, moderate efficacy against asymptomatic transmission, and strong immunogenicity in children aged 6−11 years (refer to Table 7.1). This vaccine is in use in adolescents in several countries and in February 2022 received its first approval for use in children aged 6−11 years in Australia [18].

mRNA COVID-19 vaccines have been demonstrated to be safe and well tolerated in children both in clinical trials (refer to Table 7.1), and for BNT162b2, in real-world vaccine safety surveillance data. Most local and systemic adverse events associated with these vaccines in children are mild to moderate and transient, with the most common adverse events being injection site pain, fatigue, and headache. Both mRNA vaccines are associated with a rare risk of myocarditis and/or pericarditis, and emerging evidence suggests that there is a higher risk with mRNA-1273 (Moderna) compared with BNT162b2 (Pfizer/BioNTech) [19−21]. The risk is highest for adolescent males, particularly after the second dose, and appears significantly lower in children aged 5−11 years based on preliminary data presented to the United States Advisory Committee on Immunization Practices [22]. Further research is needed to investigate this association and in particular to evaluate the risk in lower age groups.

Other COVID-19 vaccines

Inactivated and protein subunit COVID-19 vaccines have also been demonstrated to be highly immunogenic in children as young as 3 years of age (refer to Table 7.1), and the use of these platforms is increasing around the world. In clinical trials of inactivated COVID-19 vaccines in children, adverse events reported after these vaccines were mostly mild to moderate in severity and transient. Commonly reported adverse events included injection site pain, fever, fatigue, myalgia, and headache. An observational trial in Chile [23] reported a Coronavac (inactivated vaccine) vaccine effectiveness of 74% in preventing hospitalization and 90%−95% in preventing hospital and ICU admissions in children aged 6−16 years.

Ad5-nCoV (CanSinoBIO) is a viral vector vaccine that has been demonstrated to induce higher antispike and neutralizing antibody titers in children aged 6−17 years than in adults [24]; however, it has not yet been approved for use. Common adverse events reported in the clinical trial of this vaccine included injection site pain, fever, headache, and fatigue.

Over 60, phase II−IV vaccine trials are being conducted in children and adolescents with vaccine candidates across all platforms (see Table 7.2).

Correlates of protection

Vaccine-induced immunity involves multiple arms of the immune system, but for some vaccines, specific components such as antibody titers can provide a reliable correlate of clinical protection against disease, despite only representing one element of a multifaceted immune response. To date, no laboratory correlate of protection (CoP) has been established for COVID-19 vaccines. The majority of COVID-19 vaccine trials in children to date have reported only on immunogenicity rather than efficacy or effectiveness (refer to Table 7.1) and without an established CoP, these data are difficult to interpret. The relative importance of antispike antibodies, neutralizing antibodies, and cellular immune responses against SARS-CoV-2 are not yet fully understood.

The establishment of a universally accepted CoP for SARS-CoV-2 would enable more rapid development and approval of new vaccines, direct head-to-head comparison of different vaccines, and the ability to evaluate whether an individual has mounted an adequate response to a COVID-19 vaccine, guiding clinical decisions such as the need for further doses.

Two studies have demonstrated that SARS-CoV-2 neutralizing antibody is highly predictive of immune protection and correlates with vaccine efficacy [25,26] demonstrated in clinical trials. These studies were conducted before

TABLE 7.2 Current and ongoing clinical trials investigating COVID-19 vaccination in children and adolescents.

Candidate	Phase	Platform	Recruitment (no., age)	Location	Dose interval	Planned start date	Planned end date	Clinical trial record
'Com-COV3', University of Oxford Dose 1: BNT162b2 (30 μg) Dose 2: Either BNT162b2 (30 μg) *OR* BNT162b2 (10 μg) *OR* NVX-CoV2373 (5 μg or 2.5 μg) *[mRNA-1273 (50 μg) arm removed]*	II	mRNA, protein	*360,* 12–16 years	UK	≥8 weeks	September 14, 2021	December 31, 2022	ISRCTN12348322
BNT162b2 (30 μg), BioNTech/Pfizer	II/III	mRNA	*43,998,* ≥12 years	Multiple locations	21 days	April 29, 2020	May 15, 2023	NCT04368728
BNT162b2 booster dose (10 μg or 30 μg), BioNTech SE/Pfizer	III	mRNA	*10,000,* ≥12 years	Multiple locations	≥12 months after primary series for 12–17 year group	July 01, 2021	September 20, 2023	NCT04955626
BNT162b2 (30 μg), BioNTech SE/ Pfizer	II	mRNA	*420 (180 children),* ≥2 years	USA, Brazil & Germany	21 days	October 15, 2021	December 18, 2022	NCT04895982

BNT162b2, BioNTech SE/Pfizer BNT162b2 (3 μg) for 0.5 to <5 years **BNT162b2 (10 μg) for 5 to 11 years**	I/II/III	mRNA	*11,422*, 0.5 −11 years	USA	21 days	March 24, 2021	May 05, 2026	NCT04816643
BNT162b2, Pfizer	IV	mRNA	*70,000*, ≥12 years	USA	21 days	August 25, 2021	December 30, 2022	NCT05020145
BNT162b2, Rigshospitalet Denmark	IV	mRNA	*350,000*, 5 −11 years	Denmark	Not specified	December 01, 2021	April 28, 2022	NCT05186571
BNT162b2, Rabin Medical Center	IV	mRNA	*150*, 5−11 years	Israel	21 days	January 10, 2022	June 30, 2022	NCT05175989
mRNA-1273 'KidCOVE', Moderna mRNA-1273 (25 μg) for 0.5 to <2 years mRNA-1273 (25 μg or 50 μg) for 2 to <6 years **mRNA-1273 (50 μg) for 6 to <12 years**	II/III	mRNA	*13,275*, 0.5 −11 years	United States and Canada	28 days	March 15, 2021	June 12, 2023	NCT04796896

Continued

TABLE 7.2 Current and ongoing clinical trials investigating COVID-19 vaccination in children and adolescents.—cont'd

Candidate	Phase	Platform	Recruitment (no., age)	Location	Dose interval	Planned start date	Planned end date	Clinical trial record
mRNA-1273 (100 μg) 'TeenCOVE', Moderna	II/III	mRNA	*3,732*, 12–17 years	USA	28 days	December 09, 2020	April 28, 2023	NCT04649151
mRNA-1273 co-administered 9vHPV vaccine, Merck Sharp & Dohme Corp.	III	mRNA	*400*, 9–11 years	Not listed	1 month	February 02, 2022	February 22, 2023	NCT05119855
ChAdOx1-S (standard dose), University of Oxford	II	Viral vector	*300*, 6–17 years	UK	28 or 84 days	March 02, 2021	September 30, 2022	ISRCTN15638344
Ad26.COV2.S, Janssen	II	Viral vector	*635*, ≥12 years	Multiple locations	N/A	August 28, 2020	March 28, 2022	NCT04535453
Ad26.COV2.S, Janssen	II/III	Viral vector	*3,300*, 12–17 years	Multiple locations	56 days [2 dose substudy]	September 27, 2021	October 31, 2023	NCT05007080
Sputnik V, Gamelaya Research Institute	II/III	Viral vector	*3,000*, 12–17 years	Russia	21 days	July 05, 2021	November 06, 2022	NCT04954092
Ad5-nCoV, CanSino Biologics	III	Viral vector	*2,000*, 6–17 years	Mexico & Chile	56 days	February 20, 2022	August 20, 2023	NCT05169008
Covaxin, Bharat Biotech	II/III	Inactivated	*525*, 2–17 years	India	28 days	May 26, 2021	January 25, 2022	NCT04918797

CoronaVac, Sinovac	II	Inactivated	*500*, 3−17 years	China	28 days	May 03, 2021	January 03, 2022	NCT04884685
CoronaVac, Sinovac	IV	Inactivated	*2,520*, 3−17 years	China	28 days	July 27, 2021	June 25, 2022	NCT05112913
CoronaVac, Sinovac	III	Inactivated	*1,000*, 3−11 years	China	28 days	November 27, 2021	February 27, 2022	NCT05137418
CoronaVac, Sinovac	IV	Inactivated	*400*, 3−11 years	China	28 days	January 08, 2022	June 30, 2022	NCT05198336
CoronaVac, Sinovac	III	Inactivated	*14,000*, 0.5 −17 years	Multiple locations	28 days	September 10, 2021	May 25, 2022	NCT04992260
BBIBP-CorV, Sinopharm & Beijing Institute of Biological Products	III	Inactivated	*1,800*, 3−11, ≥18 years	UAE	21 or 28 days	June 06, 2021	February 06, 2022	NCT04917523
BBIBP-CorV, Sinopharm & Beijing Institute of Biological Products	IV	Inactivated	*4,400*, ≥3 years	China	21 days	April 29, 2021	March 31, 2022	NCT04863638
NVX-CoV2373, Novavax	III	Protein	*33,000*, 12 −17, ≥18 years	Multiple locations	21 days	December 27, 2020	June 30, 2023	NCT04611802
ZF2001, Anhui Zhifei Longcom Biologic Co.	I	Protein	*75*, 3−17 years	China	1 month	July 03, 2021	December 01, 2022	NCT0491359
MVC-COV1901, Medigen Vaccine Biologics Corp.	II	Protein	*399*, 12−17 years	Taiwan	28 days	July 22, 2021	April 01, 2022	NCT04951388

Continued

TABLE 7.2 Current and ongoing clinical trials investigating COVID-19 vaccination in children and adolescents.—cont'd

Candidate	Phase	Platform	Recruitment (no., age)	Location	Dose interval	Planned start date	Planned end date	Clinical trial record
BNT162b2 OR CoronaVac, University of Hong Kong	II	mRNA, inactivated	*1,000*, ≥3 years	Hong Kong	21 or 28 days	May 08, 2021	March 31, 2025	NCT04800133
BNT162b2 OR mRNA-1273 OR ChAdOx1-S, UMC Utrecht	IV	mRNA, viral vector	*640*, ≥12 years	Netherlands	Not specified	February 03, 2021	June 30, 2023	NCT05145348
Approved COVID-19 vaccines, University of British Columbia	IV	Various	*6,325*, ≥5 years	Canada	Not specified	January 01, 2022	December 01, 2025	NCT05212792
Approved mRNA COVID-19 vaccines, Duke University	IV	mRNA	*320*, 5–15 years	USA	Not specified	January 15, 2022	July 01, 2023	NCT05157191
Approved mRNA COVID-19 vaccines co-administered with IIV4 vaccine, Duke University	IV	mRNA	*450*, ≥12 years	USA	Not specified	October 04, 2021	July 14, 2022	NCT05028361

Approved COVID-19 vaccines, National Institute of Allergy and Infectious Diseases (NIAID)	IV	Various	*600*, ≥3 years	USA	Not specified	April 20, 2021	June 30, 2025	NCT04852276
COVID-19 vaccine candidates which are not routinely in use								
KCONVAC, Shenzhen Kangtai Biological Products Co., Beijing Minhai Biotechnology Co.	I	Inactivated	*84*, 3–17 years	China	28 days	August 01, 2021	October 01, 2022	NCT05003479
KCONVAC, Shenzhen Kangtai Biological Products Co., Beijing Minhai Biotechnology Co.	II	Inactivated	*480*, 3–17 years	China	28 days	September 01, 2021	April 01, 2023	NCT05003466
COVIran Barekat, Shifa Pharmed Industrial Co.	I/II	Inactivated	*500*, 12–18 years	Iran	28 days	November 11, 2021	Not listed	IRCT20171122037571N3
UB-612, Vaxxinity Inc. & United Biomedical Asia Inc.	II	Protein	*3,850*, 12–85 years	Taiwan	28 days	January 30, 2021	June 30, 2022	NCT04773067
Recombinant SARS-CoV-2 Vaccine (CHO Cell), National Vaccine and Serum Institute China	I/II	Protein	*3,580*, ≥3 years	China	30 days	April 25, 2021	October 25, 2022	NCT04869592

Continued

TABLE 7.2 Current and ongoing clinical trials investigating COVID-19 vaccination in children and adolescents.—cont'd

Candidate	Phase	Platform	Recruitment (no., age)	Location	Dose interval	Planned start date	Planned end date	Clinical trial record
SCB-2019, Clover Biopharmaceuticals	II/III	Protein	*3,820*, <18 years	Colombia	21 days	March 01, 2022	July 01, 2024	NCT05193279
BECOV2A/ Corbevax, Biological E Ltd	II/III	Protein	*624*, 5–17 years	India	28 days	October 11, 2021	August 11, 2022	CTRI/2021/10/037,066
COVAX-19/ SpikoGen, Vaxine & CinnaGen Co.	II	Protein	*610*, 12–18, 18–40 years	Iran	21 days	January 05, 2021	Not listed	IRCT20150303021315N27
ZyCoV-D, Zydus Cadila Healthcare	III	DNA	*28,216 (1000 adolescents)*, 12–18, ≥18 years	India	28 days	January 20, 2021	September 20, 2022	CTRI/2021/01/030,416
Oral bacillus subtilis spore extract COVID-19 vaccine, DreamTec Research Ltd & Hong Kong Metropolitan University	I	Spore	*30*, 12–85 years	Hong Kong	N/A	June 01, 2021	July 30, 2022	NCT05158855

Main reference used for approvals across different countries.Reuters Factbox: Countries vaccinating children against COVID-19 [December 02, 2021]: https://www.reuters.com/business/healthcare-pharmaceuticals/countries-vaccinating-children-against-covid-19-2021-06-29/.

the emergence of the Delta and Omicron variants; therefore, further research is needed to confirm whether this relationship is consistent across variants.

Passive immunization

Monoclonal antibodies have been produced to assist with passive immunization and treatment of early illness for poor responders or vaccine-ineligible individuals. These antibodies have been developed from convalescent sera of people who have recovered from SARS-CoV-1 and SARS-CoV-2. Sotrovimab is the only monoclonal antibody tested in adolescents aged 12–18 years and is used for early treatment of COVID-19 [27]. Tixagevimab and cilgavimab (Evusheld) reduced the risk of developing symptomatic COVID-19 by 77% (95% CI 45–90), compared to placebo, in adults aged ≥18 years [28]. Both monoclonal antibodies were immunogenic against the SARS-CoV-2 Omicron variant.

Adverse events following immunization

Adverse events following immunization (AEFI) continued to be monitored following clinical trials and regulatory approvals. Postlicensure (or postmarketing) surveillance is critical as conditions for vaccine use and safety monitoring can change the following licensure [29–33]. This is because clinical trials are not large enough to identify rare serious adverse events, and vaccinations are given to a broader population than those included in clinical trials. Other factors such as incorrect administration practices can also lead to AEFIs and need to be monitored.

Passive or spontaneous reporting system, where AEFIs are reported by healthcare providers and consumers to a regulatory body, remains the cornerstone of postlicensure vaccine pharmacovigilance system. In some countries, this is complemented by active surveillance systems. One such system is the electronic cohort event monitoring system that actively solicits AEFIs from individuals via SMS or email-based surveys following vaccination. AusVaxSafety (Australia) [29], ZOE Symptom Study (UK) [33], and V-Safe (US) [30–33] are examples of active vaccine safety surveillance systems in use. Another form of active surveillance is the use of large-linked databases. The Vaccine Safety Datalink project [34] is an example of this where health administrative dataset across nine healthcare organizations where health administrative data is electronically linked to vaccine records to look for rare and later onset AEFIs and investigate safety signals identified through passive/spontaneous reporting systems. The combination of safety data through passive and active vaccine surveillance systems worldwide has provided ongoing safety data for COVID vaccines used in the pediatric population.

Common or expected adverse events

BNT162b2 (Pfizer) vaccine is well tolerated in both 5–11-year-old group at the 10 μg dose and also for 12–16 year-olds at the 30 μg dose. Both phase II/III clinical trials and postlicensure safety surveillance [35,36] found most adverse events were mild and transient, usually lasting only 1–2 days [12,13]. Pain at the injection site was the most common local adverse reaction reported. A small proportion (<1%) of individuals report swollen lymph nodes (lymphadenopathy) on the side of their vaccination. Common systemic adverse events included fatigue, headache, and muscle ache. Both local and systemic adverse events in the 5–11 year age group were milder than in the older age groups [35,37]. Adverse events were more commonly reported following dose 2 compared to dose 1. Similar local and systemic adverse event profiles have been reported following mRNA-1273 (Moderna) [38] for 12–15 years compared to BNT162b2 (Pfizer).

Rare adverse events

Children with a history of anaphylaxis to polyethylene glycol, a common ingredient in other medications, hand sanitizers, cosmetics, and bathroom products, are contraindicated to have either mRNA vaccines. Reassuringly, no cases of anaphylaxis were identified in the clinical trials.

While no myocarditis or pericarditis was identified in clinical trials of either mRNA vaccine, postlicensure surveillance found an increased risk of myocarditis and pericarditis in the first 7 days following vaccination with both mRNA vaccines. Myocarditis and pericarditis can present with symptoms of chest pain, pressure or discomfort in the chest, heart palpitations or irregular heartbeats, fainting, shortness of breath, or pain with breathing. This risk is highest in adolescent males aged 12–17 years and following second dose, reported at a rate of 70–100 million doses in the United States [39]. There is emerging evidence that the risk is slightly higher following mRNA-1273 (Moderna) compared with BNT162b2 (Pfizer/BioNTech). While most individuals have had symptom resolution with conservative management, the long-term sequelae following vaccine-induced myocarditis and pericarditis remain unknown. Myocarditis from other causes can lead to arrhythmias, ventricular dysfunction, and heart failure. Therefore, individuals with myocarditis and/or pericarditis after an mRNA COVID-19 vaccine should be followed up by a specialist for at least 12 months for management, including a graded return to competitive sport/exercise, and discussion of options for further COVID-19 vaccines. Reassuringly, the risk of myocarditis and/or pericarditis appears significantly lower in children aged 5–11 years based on US data where over 15 million doses have been administered in 5–11-year-old children.

To date, there have been no reports of PIMS-TS reported following COVID vaccination in children and no evidence of menstrual disturbance in young adolescent females or fertility concerns reported post-COVID vaccination.

Vaccinating children

COVID-19 vaccines were developed at a rapid pace, with novel platforms being utilized. Vaccines have been shown to protect from severe disease, despite waning effectiveness against infection, especially against the Omicron variant. The vaccines have been shown to be safe in the pediatric population, but carry the rare risk of myocarditis, in the adolescent population. Ultimately the decision to vaccinate lies in the hands of the children and their guardians. For this, informed decision making is vital, and tools such as decision aids are important for children and their guardians to understand the risk and benefits of vaccination [40].

References

[1] Viner RM, Ward JL, Hudson LD, Ashe M, Patel SV, Hargreaves D, et al. Systematic review of reviews of symptoms and signs of COVID-19 in children and adolescents. Arch Dis Child 2021;106(8):802−7.

[2] Marks KJ. Hospitalization of infants and children aged 0−4 Years with laboratory-confirmed COVID-19—COVID-NET, 14 States, March 2020−February 2022. MMWR Morb Mortal Wkly Report. 2022;71.

[3] Roarty C, Waterfield T. Review and future directions for PIMS-TS (MIS-C). Arch Dis Child 2022. archdischild-2021-323143.

[4] Flood J, Shingleton J, Bennett E, Walker B, Amin-Chowdhury Z, Oligbu G, et al. Paediatric multisystem inflammatory syndrome temporally associated with SARS-CoV-2 (PIMS-TS): prospective, national surveillance, United Kingdom and Ireland, 2020. The Lancet Regional Health − Europe. 2021;3.

[5] Flanagan KL, MacIntyre CR, McIntyre PB, Nelson MR. SARS-CoV-2 vaccines: where are we now? J Allergy Clin Immunol Pract 2021;9(10):3535−43.

[6] McIntyre P, Joo YJ, Chiu C, Flanagan K, Macartney K. COVID-19 vaccines−are we there yet? Aust Prescr 2021;44(1):19.

[7] Koirala A, Joo YJ, Khatami A, Chiu C, Britton PN. Vaccines for COVID-19: the current state of play. Paediatr Respir Rev 2020;35:43−9.

[8] Powell AA, Kirsebom F, Stowe J, McOwat K, Saliba V, Ramsay ME, et al. Adolescent vaccination with BNT162b2 (Comirnaty, Pfizer-BioNTech) vaccine and effectiveness against COVID-19: national test-negative case-control study, England. medRxiv 2022. 2021.12.10.21267408.

[9] Reis BY, Barda N, Leshchinsky M, Kepten E, Hernán MA, Lipsitch M, et al. Effectiveness of BNT162b2 vaccine against Delta variant in adolescents. N Engl J Med 2021;385(22):2101−3.

[10] Glatman-Freedman A, Hershkovitz Y, Kaufman Z, Dichtiar R, Keinan-Boker L, Bromberg M. Effectiveness of BNT162b2 vaccine in adolescents during outbreak of SARS-CoV-2 delta variant infection, Israel, 2021. Emerg Infect Dis 2021;27(11):2919−22.

[11] Olson SM, Newhams MM, Halasa NB, Price AM, Boom JA, Sahni LC, et al. Effectiveness of BNT162b2 vaccine against critical Covid-19 in adolescents. N Engl J Med 2022;386(8):713−23.

[12] Walter EB, Talaat KR, Sabharwal C, Gurtman A, Lockhart S, Paulsen GC, et al. Evaluation of the BNT162b2 Covid-19 vaccine in children 5 to 11 years of age. N Engl J Med 2021;386(1):35–46.

[13] Frenck RW, Klein NP, Kitchin N, Gurtman A, Absalon J, Lockhart S, et al. Safety, immunogenicity, and efficacy of the BNT162b2 Covid-19 vaccine in adolescents. N Engl J Med 2021;385(3):239–50.

[14] Thomas SJ, Moreira ED, Kitchin N, Absalon J, Gurtman A, Lockhart S, et al. Safety and efficacy of the BNT162b2 mRNA Covid-19 vaccine through 6 months. N Engl J Med 2021;385(19):1761–73.

[15] Dorabawila V, Hoefer D, Bauer UE, Bassett MT, Lutterloh E, Rosenberg ES. Effectiveness of the BNT162b2 vaccine among children 5-11 and 12-17 years in New York after the emergence of the Omicron variant. medRxiv 2022. 2022.02.25.22271454.

[16] Fowlkes AL, Yoon SK, Lutrick K, Gwynn L, Burns J, Grant L, et al. Effectiveness of 2-dose BNT162b2 (Pfizer BioNTech) mRNA vaccine in preventing SARS-CoV-2 infection among children aged 5–11 years and adolescents aged 12–15 years—PROTECT Cohort, July 2021–February 2022.

[17] Pfizer and BioNTech provide update on ongoing studies of COVID-19 vaccine [press release]. Online, December 17, 2021.

[18] Therapeutic Goods Administration. COVID-19 vaccine: SPIKEVAX (elasomeran). 2021 [updated February 23, 2022]. Available from: https://www.tga.gov.au/covid-19-vaccine-spikevax-elasomeran.

[19] Klein NP. Myocarditis analyses in the vaccine safety datalink rapid cycle analyses and "Head-to-head" product comparison. 2021. Available from: https://www.cdc.gov/vaccines/acip/meetings/downloads/slides-2021-10-20-21/08-COVID-Klein-508.pdf.

[20] Wong H-L. Surveillance updates of myocarditis/pericarditis and mRNA COVID-19 vaccination in the FDA BEST system: US Food & Drug administration. 2021 updated October 14, 2021. Available from: https://www.fda.gov/media/153090/download.

[21] Agency MaHR. Coronavirus vaccine - weekly summary of Yellow Card reporting. 2021 [updated March 17, 2022]. Available from: https://www.gov.uk/government/publications/coronavirus-covid-19-vaccine-adverse-reactions/coronavirus-vaccine-summary-of-yellow-card-reporting.

[22] Su JR. COVID-19 vaccine safety updates: Primary series in children and adolescents ages 5–11 and 12–15 years, and booster doses in adolescents ages 16–24 years. 2022 [updated January 5, 2022]. Available from: https://www.cdc.gov/vaccines/acip/meetings/downloads/slides-2022-01-05/02-COVID-Su-508.pdf.

[23] Jara A, Undurraga EA, Flores JC, Zubizarreta JR, González C, Pizarro A, et al. Effectiveness of an inactivated SARS-CoV-2 vaccine in children and adolescents: a large-scale observational study. SSRN: preprints with the Lancet. 2022.

[24] Zhu F, Jin P, Zhu T, Wang W, Ye H, Pan H, et al. Safety and immunogenicity of a recombinant adenovirus type-5–vectored coronavirus disease 2019 (COVID-19) vaccine with a homologous prime-boost regimen in healthy participants aged ≥6 Years: a randomized, double-blind, placebo-controlled, phase 2b trial. Clin Infect Dis 2021.

[25] Khoury DS, Cromer D, Reynaldi A, Schlub TE, Wheatley AK, Juno JA, et al. Neutralizing antibody levels are highly predictive of immune protection from symptomatic SARS-CoV-2 infection. Nat Med 2021;27(7):1205–11.

[26] Earle KA, Ambrosino DM, Fiore-Gartland A, Goldblatt D, Gilbert PB, Siber GR, et al. Evidence for antibody as a protective correlate for COVID-19 vaccines. Vaccine 2021;39(32):4423–8.

[27] GlaxoSmithKline. Primary endpoint met in COMET-TAIL phase III trial evaluating intramuscular administration of sotrovimab for early treatment of COVID-19. 2021 [updated November 12, 2021. Available from: https://www.gsk.com/en-gb/media/press-releases/primary-endpoint-met-in-comet-tail-phase-iii-trial-evaluating-intramuscular-administration-of-sotrovimab-for-early-treatment-of-covid-19/.

[28] AstraZeneca. AZD7442 PROVENT Phase III prophylaxis trial met primary endpoint in preventing COVID-19; 2021 [updated 20 August 2021].

[29] Phillips A, Carlson S, Danchin M, Beard F, Macartney K. From program suspension to the pandemic: a qualitative examination of Australia's vaccine pharmacovigilance system over 10 years. Vaccine 2021;39(40):5968–81.

[30] Gee J, Marquez P, Su J, Calvert GM, Liu R, Myers T, et al. First month of COVID-19 vaccine safety monitoring — United States, December 14, 2020–January 13, 2021. MMWR Morb Mortal Wkly Rep 2021;70(8):283–8.

[31] Chen G, Li X, Sun M, Zhou Y, Yin M, Zhao B, et al. COVID-19 mRNA vaccines are generally safe in the short term: a vaccine vigilance real-world study says. Front Immunol 2021;12.

[32] Shimabukuro TT, Kim SY, Myers TR, Moro PL, Oduyebo T, Panagiotakopoulos L, et al. Preliminary findings of mRNA Covid-19 vaccine safety in pregnant persons. N Engl J Med 2021;384(24):2273–82.

[33] Menni C, Klaser K, May A, Polidori L, Capdevila J, Louca P, et al. Vaccine side-effects and SARS-CoV-2 infection after vaccination in users of the COVID Symptom Study app in the UK: a prospective observational study. Lancet Infect Dis 2021;21(7):939–49.

[34] Chen RT, Glasser JW, Rhodes PH, Davis RL, Barlow WE, Thompson RS, et al. Vaccine Safety Datalink project: a new tool for improving vaccine safety monitoring in the United States. The Vaccine Safety Datalink Team. Pediatrics 1997;99(6):765–73.

[35] AusVaxSafety. COVID-19 vaccines. 2022 [updated March 21, 2022]. Available from: https://ausvaxsafety.org.au/safety-data/covid-19-vaccines.

[36] Hause AM, Baggs J, Marquez P, Myers TR, Gee J, Su JR, et al. COVID-19 vaccine safety in children aged 5–11 Years—United States, November 3–December 19, 2021. MMWR (Morb Mortal Wkly Rep) 2021;70(51–52):1755–60.

[37] United States Food and Drug Administration. Vaccines and related biological products advisory committee October 26, 2021 Meeting Document. Maryland2021.

[38] Ali K, Berman G, Zhou H, Deng W, Faughnan V, Coronado-Voges M, et al. Evaluation of mRNA-1273 SARS-CoV-2 vaccine in adolescents. N Engl J Med 2021;385(24):2241–51.

[39] Oster ME, Shay DK, Su JR, Gee J, Creech CB, Broder KR, et al. Myocarditis cases reported after mRNA-based COVID-19 vaccination in the US from December 2020 to august 2021. JAMA 2022;327(4):331–40.

[40] Frawley J, Wiley K, Leask J, Danchin M, Mahimbo A, Lyndal T, et al. Decision aid (5–15 years): should I get the COVID-19 vaccine for my child? 2022 updated Feb 2022. Available from: https://www.ncirs.org.au/covid-19-decision-aid-for-children.

[41] Australian Government Department of Health Therapeutic Goods Administration. Pfizer's COVID-19 vaccine (COMIRNATY) provisionally approved for use in individuals 5 years and over. 2021. Available from: https://www.tga.gov.au/media-release/pfizers-covid-19-vaccine-comirnaty-provisionally-approved-use-individuals-5-years-and-over.

[42] European Medicines Agency. Comirnaty COVID-19 vaccine: EMA recommends approval for children aged 5 to 11. 2021. Available from: https://www.ema.europa.eu/en/news/comirnaty-covid-19-vaccine-ema-recommends-approval-children-aged-5-11.

[43] United Kingdom Government Medicines and Healthcare products Regulatory Agency. UK regulator approves use of Pfizer/BioNTech vaccine in 5 to 11-year olds. 2021. Available from: https://www.gov.uk/government/news/uk-regulator-approves-use-of-pfizerbiontech-vaccine-in-5-to-11-year-olds.

[44] United States Food and Drug Administration. FDA Authorizes Pfizer-BioNTech COVID-19 Vaccine for Emergency Use in Children 5 through 11 years of age. 2021. Available from: https://www.fda.gov/news-events/press-announcements/fda-authorizes-pfizer-biontech-covid-19-vaccine-emergency-use-children-5-through-11-years-age.

[45] Health Canada. Health Canada authorizes use of Comirnaty (the Pfizer-BioNTech COVID-19 vaccine) in children 5 to 11 years of age. 2021. Available from: https://www.canada.ca/en/health-canada/news/2021/11/health-canada-authorizes-use-of-comirnaty-the-pfizer-biontech-covid-19-vaccine-in-children-5-to-11-years-of-age.html.

[46] New Zealand Medsafe. Approval status of COVID-19 vaccines applications received by Medsafe. 2021. Available from: https://www.medsafe.govt.nz/COVID-19/status-of-applications.asp.

[47] Government of Mexico COFEPRIS. COFEPRIS issues modification to the authorization for emergency use of the Pfizer-BioNTech vaccine will allow application from 12 years old. 2021. Available from: https://www.gob.mx/cofepris/articulos/cofepris-emite-modificacion-a-la-autorizacion-para-uso-de-emergencia-de-vacuna-pfizer-biontech-permitira-aplicacion-a-partir-de-12-anos?idiom=es.

[48] Morland S, Baldassari V, Cherfan O, Portala J, Davis C, Marchioro L. Factbox: countries vaccinating children against COVID-19. Reuters. December 2, 2021.

[49] United Arab Emirates Ministry of Health and Prevention. COVID-19 vaccine for children. 2021. Available from: https://u.ae/en/information-and-services/justice-safety-and-the-law/handling-the-covid-19-outbreak/vaccines-against-covid-19-in-the-uae/covid19-vaccine-for-children.

[50] Araujo da Silva A, de Carvalho B, Esteves M, Teixeira C, Souza C. Role of COVID-19 vaccinal status in admitted children during OMICRON variant circulation in Rio de Janeiro, city- Preliminary report. medRxiv 2022. 2022.02.10.22270817.

[51] Burns MD, Boribong BP, Bartsch YC, Loiselle M, Davis JP, Lima R, et al. Durability and cross-reactivity of SARS-CoV-2 mRNA vaccine in adolescent children. medRxiv 2022. 2022.01.05.22268617.

[52] Lutrick K, Rivers P, Yoo YM, Grant L, Hollister J, Jovel K, et al. Interim estimate of vaccine effectiveness of BNT162b2 (Pfizer-BioNTech) vaccine in preventing SARS-CoV-2 infection among adolescents aged 12−17 years — Arizona, July−December 2021. MMWR Morb Mortal Wkly Report 2021;70(51−52):1761−5.

[53] Mevorach D, Anis E, Cedar N, Hasin T, Bromberg M, Goldberg L, et al. Myocarditis after BNT162b2 vaccination in Israeli adolescents. N Engl J Med 2022;386:998−9.

[54] Moderna's COVID-19 vaccine (SPIKEVAX) provisionally approved for use in individuals 6 years and older [press release]. Online, February 17, 2022.

[55] COVID-19 vaccine Spikevax approved for children aged 12 to 17 in EU [press release]. Online, July 23, 2021.

[56] Moderna COVID-19 vaccine approved by MHRA in 12-17 year olds [press release]. Online, August 17, 2021.

[57] Health Canada. Vaccines for children: COVID-19 online 2021 [updated 28 January 2022]. Available from: https://www.canada.ca/en/public-health/services/vaccination-children/covid-19.html.

[58] Swissmedic approves indication extension for Spikevax vaccine for 12- to 17-year-olds [press release]. Online, August 9, 2021.
[59] Invima approves the use of Moderna's vaccine for children under 18 years of age. El Colombiano. September 20, 2021.
[60] Moderna reports third quarter fiscal year 2021 financial results and provides business updates [press release]. Online, November 4, 2021.
[61] Moderna announces positive top line data from phase 2/3 study of COVID-19 vaccine in children 6 to 11 years of age [press release]. Online, October 25, 2021.
[62] Therapeutic Goods Administration. Australian public assessment report for Elasomeran. In: Health AGDo; 2022. p. 1−65. online.
[63] Bartsch YC, St Denis KJ, Kaplonek P, Kang J, Lam EC, Burns MD, et al. Comprehensive antibody profiling of mRNA vaccination in children. bioRxiv. 2022:2021.10.07.463592.
[64] RDIF ready to provide the one-shot Sputnik Light vaccine and Sputnik M vaccine (for adolescents) as it develops strategic cooperation with partners in India to fight COVID [press release]. Online, December 6, 2021.
[65] Kazakhstan becomes the first country outside Russia to authorize Sputnik M vaccine for adolescents [press release]. Online, February 22, 2022.
[66] TASS. Efficacy of Sputnik M vaccine against coronavirus amounts to 93%, web portal says. TASS Russian News Agency. January 11, 2022.
[67] Zhu J. Hong Kong authorises Sinovac vaccine for children aged 3-17. Reuters. November 20, 2021.
[68] The City Paper Staff. Colombia starts COVID-19 vaccination campaign for children ages 3 to 11. The City Paper Bogota. November 2, 2021.
[69] Han B, Song Y, Li C, Yang W, Ma Q, Jiang Z, et al. Safety, tolerability, and immunogenicity of an inactivated SARS-CoV-2 vaccine (CoronaVac) in healthy children and adolescents: a double-blind, randomised, controlled, phase 1/2 clinical trial. Lancet Infect Dis 2021;21(12):1645−53.
[70] Soto JA, Melo-González F, Gutierrez-Vera C, Schultz BM, Berríos-Rojas RV, Rivera-Pérez D, et al. An inactivated SARS-CoV-2 vaccine is safe and induces humoral and cellular immunity against virus variants in healthy children and adolescents in Chile. medRxiv 2022. 2022.02.15.22270973.
[71] Xia S, Zhang Y, Wang Y, Wang H, Yang Y, Gao GF, et al. Safety and immunogenicity of an inactivated COVID-19 vaccine, BBIBP-CorV, in people younger than 18 years: a randomised, double-blind, controlled, phase 1/2 trial. Lancet Infect Dis 2021;22(2):196−208.
[72] Tawinprai K, Siripongboonsitti T, Porntharukchareon T, Vanichsetakul P, Thonginnetra S, Niemsorn K, et al. Safety and immunogenicity of the BBIBP-CorV vaccine in adolescents aged 12-17 years in Thai population, prospective cohort study. medRxiv 2022. 2022.01.07.22268883.
[73] Interim statement on COVID-19 vaccination for children and adolescents [press release]. Online, November 24, 2021.
[74] Vadrevu KM, Reddy S, Jogdand H, Ganneru B, Mirza N, Tripathy VN, et al. Immunogenicity and safety of an inactivated SARS-CoV-2 vaccine (BBV152) in children from 2 to 18 years of age: an open-label, age-de-escalation phase 2/3 study. medRxiv 2021. 2021.12.28.21268468.
[75] Novavax announces positive results of COVID-19 vaccine in pediatric population of PREVENT-19 phase 3 clinical trial [press release]. Online, February 10, 2022.
[76] The CECMED approves the Emergency Use Authorization for the Cuban SOBERANA® 02 vaccine in the pediatric population [press release]. Online, September 3, 2021.

[77] Sequera V, Kinosian S. Venezuela begins vaccinating 2-year-old children with Cuban doses -vice president. Reuters. November 9, 2021.
[78] The CECMED approves the Emergency Use Authorization of the Cuban vaccine candidate ABDALA [press release]. Online, July 9, 2021.
[79] Telesur. Cuba: abdala vaccine authorized in children from 2-11 Years old. Telesur. October 28, 2021.

Chapter 8

Long COVID in children

Joseph L. Mathew and Kamal Kumar Singhal
Department of Pediatrics, Advanced Pediatrics Centre, Postgraduate Institute of Medical Education and Research (PGIMER), Chandigarh, India

Introduction

Coronaviruses are RNA viruses with considerable genetic diversity, that enables them to jump across the species barrier. In late 2019, a novel coronavirus emerged in Wuhan City of Hubei Province of China, causing an outbreak of an unusual respiratory illness, leading to severe pneumonia and cardiorespiratory compromise in infected persons. The International Committee on Taxonomy of Viruses named it Severe Acute Respiratory Syndrome Coronavirus 2 (SARS-CoV-2) [1]. The illness associated with it was named Coronavirus disease 2019 (abbreviated as COVID-19) by the World Health Organization (WHO). By the end of December 2021, globally, more than 375 million people were reported to be affected by COVID-19, with fatal outcomes in almost 5.6 million [2]. The American Academy of Pediatrics reported that more than seven million children were infected by the end of December 2021, representing 17.3% of the total cases in the United States [3]. Compared to adults, children with COVID-19 are typically asymptomatic or experience mild symptoms, with survival ranging from 97% to 99.9% [3,4]. Among all hospitalized cases of COVID-19, children contribute only 0.27% [3]. Similarly, among all COVID deaths, pediatric cases account for just 0.3% [3]. However, recently concerns have been raised regarding the persistence of signs and symptoms beyond the usual duration of the acute illness. This is referred to as "Long COVID" and is better described in adult patients. However, the prevalence and characterization of Long COVID in children are less well known.

What is long COVID?

As healthcare workers and various public health professionals continue to understand the evolving COVID-19 pandemic, there is a need to define and characterize the symptoms people are experiencing. Various healthcare institutions and organizations have attempted to evolve a definition and criteria

Clinical Management of Pediatric COVID-19. https://doi.org/10.1016/B978-0-323-95059-6.00010-3

for Long COVID. Most of these are based on consensus among experts, and some are still evolving.

The WHO conducted a global Delphi consensus process that comprised various stakeholders, including patients, patient researchers, external experts, and their own staff, to develop a working ‘case definition’ of the post-COVID-19 state in adults [5]. They described it as a “condition in individuals with a history of probable or confirmed SARS-CoV-2 infection, usually 3 months from the onset of COVID-19, with symptoms that last for at least 2 months, that cannot be explained by an alternative diagnosis. These symptoms include fatigue, shortness of breath, cognitive dysfunction, and various others, that have an impact on everyday functioning. The symptoms may be of new onset following initial recovery from an acute COVID-19 episode or may persist from the initial illness. Symptoms may also fluctuate or relapse over time.” They further added that a separate definition for children might be applicable [5].

It is essential to note the difference between prolonged acute COVID-19 and long COVID-19. Acute COVID-19 has been defined by several guidelines as the persistence of signs and symptoms of COVID-19 for up to 4 weeks. The persistence of signs and symptoms from 4 to 12 weeks has been described as Ongoing symptomatic COVID-19 [4–7]. Long COVID has typically been defined as the persistence of clinical signs and symptoms (unexplained by another diagnosis) beyond 12 weeks after the onset of acute COVID-19 [4,6,7]. In a broader sense, Long COVID has been considered as a non-return to the usual state of health after COVID-19 illness or development of new or recurrent symptoms after the resolution of acute COVID [6–9].

Various other terms have been used to describe the long-term sequelae after acute COVID-19 infections. These include “long-haulers,” “chronic COVID,” “postacute sequelae,” “postacute sequelae of SARS-CoV-2 (PASC),” and “postacute COVID syndrome” [2,5,7].

Epidemiology

Adults: Most of the literature regarding Long COVID in adults is derived from patients hospitalized during acute illness. Among them, the prevalence of Long COVID has been reported as high as 87.4% [10]. In one study that examined the frequency of persistent symptoms among people with varying severity of acute COVID, the prevalence among those with severe disease was 89%, compared to 75% among those with disease of moderate severity, and 59% among those with a mild acute episode [11]. In adults, Long COVID consists of multisystem disease, and symptoms include fatigue, muscular weakness, shortness of breath, sleeping difficulty, depression, and anxiety [12].

Children: The long-term sequelae of acute COVID-19 in children are largely unknown. This may be due to limited awareness of the condition among children, under-reporting, and the fact that most pediatric COVID-19 cases are either mild or asymptomatic, and hence do not require hospitalization. A few studies followed patients after acute COVID for variable periods ranging from less than 30 days, to 9 months [9,13–18]. Although the precise proportion of children developing Long COVID and the predisposing risk factors are unknown, it was believed to range from 0% to 66% [9,13–18]. Recent data from Europe and Russia suggested that in children, signs and symptoms can persist beyond 3 months after the resolution of acute COVID and are similar to those in adults [19,20]. Conversely, Denina et al., who followed hospitalized children for 4 months after discharge, did not find evidence of Long COVID despite clinical, laboratory, and lung ultrasonography evaluation [21]. These divergent data highlight the marked variation both in the prevalence of persistent symptoms following an acute episode, as well as its duration.

Pathogenesis

Our understanding of the pathophysiological mechanisms of Long COVID is still evolving. Suggested mechanisms include consequences of direct virus-induced damage, host-induced inflammatory responses, and sequelae related to post-critical illness [9].

In the lungs, direct virus-induced damage and host inflammatory responses during acute illness results in the breakdown of the endothelial–epithelial barrier. This results in alveolar invasion by monocytes and neutrophils and extravasation of protein-rich exudate into the alveolar space. This diffuse alveolar damage is consistent with acute respiratory distress syndrome seen in other etiologies. COVID patients have also been found to have lung vasculature microangiopathy and macrothrombosis [9]. Accelerated lung fibrosis similar to end-stage pulmonary fibrosis has also been reported following COVID in some individuals [22]. These pathophysiologic changes often result in the persistence of pulmonary symptoms and diminished respiratory reserve, contributing to the pulmonary manifestations of Long COVID.

In contrast, the basis for various neuropsychiatric manifestations in COVID-19 is more complex. It includes overlapping pathological mechanisms such as direct viral infection, severe host inflammatory response, neurodegeneration, neuroinflammation, and microvascular thrombosis [9]. The persisting manifestations of Long COVID may be the result of chronic low-level brain inflammation (referred to as "inflammaging"), reducing the ability of the host to respond to new antigens and causing an accumulation of memory T cells [23]. The severity of the preceding acute illness influences the mechanisms and manifestations of neuropsychiatric sequelae. For example, brain fog following critically acute illness may be due to posttraumatic stress disorder and deconditioning. In contrast, dysautonomia is observed after mild acute illness [9,24].

Clinical manifestations

Long COVID involves multiple systems and may encompass physical, cognitive, and mood symptoms. The clinical manifestations of Long COVID are a cluster of overlapping signs and symptoms that may fluctuate over time. The wide variation in the clinical presentation suggests a multifactorial basis for Long COVID.

Zimmermann et al. reviewed 14 studies investigating over 19,000 children and adolescents for Long COVID [18]. All the studies were from high-income countries, with heterogeneity in study design, inclusion criteria, outcome variables, follow-up duration and time points. Almost all had major limitations, and only 5 (36.7%) studies included comparison groups. Among these five studies, only three reported an increased prevalence of persistent symptoms, among cases (compared to controls without prior COVID infection). Most of the studies used online questionnaires or telephone interviews. The most reported symptoms were fatigue (3%−87%), headache (3%−80%), concentration difficulties (2%−81%), abdominal pain (1%−76%), sleep disturbance (2%−63%), myalgia or arthralgia (1%−61%), cough (1%−30%), chest tightness or pain (1%−31%), loss of appetite or weight (2%−50%), rash (2%−52%), disturbed smell or anosmia (3%−26%), and congested or runny nose (1%−12%).

A Dutch study invited all the pediatricians in the Netherlands to share their experiences with Long COVID in children via an online survey [2]. The investigators provided a list of symptoms of Long COVID frequently reported in adults, including persistent tiredness, headaches, dyspnea, concentration problems, depression, dermatological changes, and gastrointestinal manifestations, persisting for several months after COVID infection. Participating pediatricians reviewed their patient records to identify children with the same symptoms. Complete responses were received from 78% of the pediatricians. A total of 89 children (mean age 13 years) had symptoms associated with Long COVID. The most frequent complaints were fatigue (87%), dyspnea (55%), and concentration difficulty (45%). Headache (38%), chest pain (33%), myalgia (28%), diarrhea (24%), and memory loss (13%) were also described. Brain fog, weight loss, diminished appetite, and persistent fever were each reported in approximately 2% of the patients. Only one patient had anosmia and dysgeusia. The symptoms severely limited the daily functioning in 36% of the children.

However, only 6 of the 89 patients fulfilled the criteria for Long COVID. All six reported shortness of breath and fatigue (with extreme fatigue in three). Three children had asthma-like symptoms with variable responses to bronchodilator medications. Four children reported chest pain and headache. Dizziness was reported by three children and cognitive dysfunction, cough, palpitation, abdominal pain, and change in bowel habits were reported by two children. Weight loss, nausea, fever, and skin rash were reported by one child each. Interestingly, none of these six children had required hospitalization for acute COVID. These data suggest a wide spectrum of persistent manifestations in the pediatric age group.

Morrow et al. published a case series of nine patients aged less than 19 years old, having symptoms associated with Long COVID [12]. The most frequently reported symptoms were fatigue (89%), headaches (67%), difficulty with school work (67%), brain fog (44%), and dizziness or light-headedness (44%). Some patients reported depression and anxiety symptoms [12].

Based on parental reporting, Ludvigsson et al. reported a case series of five children from Sweden who had persistent symptoms for 6−8 months [19]. These included fatigue, dyspnea, palpitations or chest pain, headaches, poor concentration, muscle weakness, dizziness, and sore throat. Three children reported persistent abdominal pain, memory loss, skin rash, depression, and muscle aches. Two children reported sleep disorders, remitting fever, diarrhea, joint pain, hyper-anesthesia, and vomiting. Due to these symptoms, they were not able to attend school for several months after the resolution of acute COVID [19].

Fink et al. compared demographic, anthropometric, and health-related quality of life characteristics among 53 post-COVID patients (8−18 years old) with 52 negative controls (matched for age, sex, and preexisting chronic condition). The controls consisted of children presenting to outpatient clinics for routine visits, children without warning signs of pediatric COVID-19, and children who were negative for SARS-CoV-2 through polymerase chain reaction and serological tests. The investigators used the PedsQL4.0 tool that includes 23 items across four domains related to physical, emotional, social, and school functioning. The median duration of follow-up was 4.4 months. The investigators defined Long COVID as the persistence of clinical abnormalities beyond 12 weeks of the onset of acute COVID-19 that could not be explained by other conditions. Although approximately 40% of the patients had at least one persistent symptom on follow-up, only 23% had Long COVID. The most common symptoms were headache (19%), severe recurrent headache (9%), tiredness (9%), dyspnea (8%), and concentration difficulty (4%). Compared to the controls, the PedsQL4.0 scores of post-COVID patients were lower in the domains related to physical and school functioning but comparable in the other two domains [4].

Very recently, data from the LongCOVIDKidsDK study became available [25]. This was a nationwide survey of adolescents (15−18 years old) in Denmark, comparing the prevalence of symptoms (associated with Long COVID in adults) among those with previous COVID infection (i.e., cases), and age- and gender-matched controls who did not have prior COVID. The investigators found that among 23 sentinel symptoms, only nine were more frequent cases. These included breathing difficulty, cough, chest pain, headache, sore throat, palpitations, dizziness, appetite loss, and dizziness upon standing. In contrast, eight symptoms were more prevalent among controls;

these included abdominal pain, rash, mood swings, cold hands/feet, discolored fingers or toes, extreme pallor, chapped lips, and dark circles under the eyes. The prevalence of six symptoms, often associated with Long COVID in adults did not show statistically significant differences between cases and controls; these included muscle/joint pain, fatigue, nausea, fever, concentration or memory problems, and photosensitivity. Further, cases had superior scores (compared to controls) for quality of life, physical functioning, emotional functioning, social functioning, and school functioning [25].

The diverse data from various studies demand explanation. It is well recognized that the pandemic and its impact on quality of life have affected the emotional and behavioral health of children and their parents [26]. It is likely that many of the reported symptoms, especially those related to mental, emotional, and physical health, could be due to the conditions associated with the pandemic rather than COVID itself. As most studies lacked comparison groups, it is difficult to distinguish the impact of the acute infection from the non-clinical effects of the pandemic. Several studies also reported low response rates from the patients who were requested to participate in the follow-up surveys. This could result in an over-estimation of the prevalence (as those with persistent symptoms are more likely to respond than those with mild or no symptoms). None of the studies correlated the initial acute disease severity with Long COVID symptoms [18].

Association of long COVID syndrome with the severity of acute illness

So far, no association has been found between the incidence or severity of Long COVID syndrome with either the severity of acute infection or any underlying chronic medical illness [27]. Long COVID may be a significant concern even in children in whom the acute COVID episode was not severe enough to merit hospitalization [18].

Diagnosis and differential diagnosis

There are currently no investigations to confirm a diagnosis of Long COVID. Symptoms described due to post-COVID conditions may have other explanations. These include pandemic-related social isolation measures, other clinical diagnoses, unmasking of pre-existing health conditions, or a new COVID-19 infection. Therefore, ruling out these confounders is imperative before diagnosing Long COVID.

Treatment

As Long COVID is a multisystem syndrome, there is a need for a multidisciplinary approach to its management [6,7]. This includes providing the patients and their caregivers with appropriate education, reliable information, symptom management strategies, physical rehabilitation, and psychological support [6]. Given the predominance of physical and mental issues in pediatric Long COVID, rehabilitation should focus on increasing physical activity and improving mental health [4,28,29]. Several experts have suggested establishing multidisciplinary rehabilitation clinics for Long COVID. The initial contact with the healthcare system should include a holistic evaluation of the patient comprising physical, cognitive, psychological, psychiatric, and functional domains. All children with acute COVID-19 infections should have at least one follow-up [30]. Some consultations may be done using telemedicine platforms as well.

At present, there is insufficient evidence to suggest there are specific pharmacological treatment options for Long COVID, although symptomatic relief of symptoms should be tried.

Patients should be provided with appropriate information related to their clinical condition and shared decision-making should be attempted [6]. Self-management should be encouraged and realistic goals should be identified in consultation with the patients and their caregivers.

Patients should also be encouraged to take COVID-19 vaccines as per the national policy to reduce the risk of new acute COVID infection [6]. Although the effects of Long COVID are still not apparent, limited evidence suggests a lower risk of Long COVID among patients who received two doses of vaccination [31].

There is currently no evidence reporting the efficacy of over-the-counter vitamins and supplements in treating Long COVID.

Some manifestations of Long COVID merit urgent referral. Severe psychiatric manifestations including increased risk of suicide or self-harm should be urgently referred to a psychiatrist. Other mood disorders or psychiatric manifestations should also be managed appropriately. Acute or life-threatening complications of Long COVID should be identified, and urgently referred to relevant acute services [6]. These complications may include hypoxemia or oxygen desaturation on exertion, signs of severe lung disease, and cardiac chest pain.

Prognosis

Although the recovery time varies from patient to patient and from symptom to symptom, most of the children and adolescents have satisfactory physical and functional recovery over 4 months of follow-up [4,6]. This is in contrast to adult patients where at least one symptom was present in 68% and 49% of the patients at 6 and 12 months of follow-up [13].

Summary

Although children with COVID-19 generally have mild symptoms and overall better prognoses than adults, some children have persistent symptoms far beyond the usual duration of the acute illness. Long COVID refers to a syndrome with the persistence of multisystem symptoms beyond 3 months of the onset of critical illness, in which other diagnoses have been ruled out. There is no association between the incidence of Long COVID, and the severity of the acute COVID infection or the presence of an underlying medical illness. Similarly, the clinical characteristics of the acute infection do not influence the severity of Long COVID symptoms. There is a paucity of well-designed studies on Long COVID in children, and there are no data from developing countries. Most of the available epidemiological data is based on heterogeneous studies, which often lacked comparison groups and collected information using online questionnaires or telephonic communication. These reports indicated a wide variation in the prevalence of Long COVID in children. The clinical manifestations of Long COVID in children resemble those in adults. The most frequently reported symptoms are headache, fatigue, sleep disturbance, concentration difficulties, abdominal pain, myalgia or arthralgia, congested or runny nose, cough, chest tightness or pain, loss of appetite or weight, disturbed smell or anosmia, and rash. In some children, these symptoms could be severe enough to limit activities of daily living. The pathophysiological mechanisms for Long COVID are poorly understood. Pandemic-associated factors such as social isolation, alternate diagnoses, and unmasking of preexisting health conditions could be possible explanations in some children. There is no confirmatory diagnostic test and no specific treatment

References

[1] Coronaviridae Study Group of the International Committee on Taxonomy of Viruses. The species Severe acute respiratory syndrome-related coronavirus: classifying 2019-nCoV and naming it SARS-CoV-2. Nat Microbiol 2020;5(4):536–44.

[2] Brackel CLH, Lap CR, Buddingh EP, van Houten MA, van der Sande LJTM, Langereis EJ, et al. Pediatric long-COVID: an overlooked phenomenon? Pediatr Pulmonol August 2021;56(8):2495–502.

[3] American Academy of Pediatrics. Children and COVID-19: state-level data report [Internet]. Home. 2021 [cited 2021 Dec 24]. Available from: http://www.aap.org/en/pages/2019-novel-coronavirus-covid-19-infections/children-and-covid-19-state-level-data-report/.

[4] Fink TT, Marques HHS, Gualano B, Lindoso L, Bain V, Astley C, et al. Persistent symptoms and decreased health-related quality of life after symptomatic pediatric COVID-19: a prospective study in a Latin American tertiary hospital. Clinics (Sao Paulo). 2021;76:e3511.

[5] World Health Organization. A clinical case definition of post COVID-19 condition by a Delphi consensus [Internet]. 2021 [cited 2022 Jan 4]. Available from: https://www.who.int/publications-detail-redirect/WHO-2019-nCoV-Post_COVID-19_condition-Clinical_case_definition-2021.1.

[6] NICE. COVID-19 rapid guideline: managing the long-term effects of COVID-19 [Internet]. NICE; 2021 [cited 2021 Dec 30]. Available from: https://www.nice.org.uk/guidance/ng188/chapter/Recommendations.

[7] CDC. Post-COVID conditions: information for healthcare providers [Internet]. Centers for Disease Control and Prevention; 2020 [cited 2021 Dec 30]. Available from: https://www.cdc.gov/coronavirus/2019-ncov/hcp/clinical-care/post-covid-conditions.html.

[8] The Lancet null. Facing up to long COVID. Lancet December 12, 2020;396(10266):1861.

[9] Nalbandian A, Sehgal K, Gupta A, Madhavan MV, McGroder C, Stevens JS, et al. Post-acute COVID-19 syndrome. Nat Med April 2021;27(4):601−15.

[10] Carfì A, Bernabei R, Landi F. Gemelli against COVID-19 post-acute care study group. Persistent symptoms in patients after acute COVID-19. JAMA August 11, 2020;324(6):603−5.

[11] Arnold DT, Hamilton FW, Milne A, Morley AJ, Viner J, Attwood M, et al. Patient outcomes after hospitalisation with COVID-19 and implications for follow-up: results from a prospective UK cohort. Thorax April 2021;76(4):399−401.

[12] Morrow AK, Ng R, Vargas G, Jashar DT, Henning E, Stinson N, et al. Postacute/long COVID in pediatrics: development of a multidisciplinary rehabilitation clinic and preliminary case series. Am J Phys Med Rehabil December 1, 2021;100(12):1140−7.

[13] Huang L, Yao Q, Gu X, Wang Q, Ren L, Wang Y, et al. 1-year outcomes in hospital survivors with COVID-19: a longitudinal cohort study. Lancet 2021;398(10302):747−58.

[14] Buonsenso D, Munblit D, De Rose C, Sinatti D, Ricchiuto A, Carfi A, et al. Preliminary evidence on long COVID in children. Acta Paediatr July 2021;110(7):2208−11.

[15] Say D, Crawford N, McNab S, Wurzel D, Steer A, Tosif S. Post-acute COVID-19 outcomes in children with mild and asymptomatic disease. Lancet Child Adolesc Health June 2021;5(6):e22−3.

[16] Radtke T, Ulyte A, Puhan MA, Kriemler S. Long-term symptoms after SARS-CoV-2 infection in children and adolescents. JAMA September 7, 2021;326(9):869−71.

[17] Farooqi KM, Chan A, Weller RJ, Mi J, Jiang P, Abrahams E, et al. Longitudinal outcomes for multisystem inflammatory syndrome in children. Pediatrics August 2021;148(2). e2021051155.

[18] Zimmermann P, Pittet LF, Curtis N. How common is long COVID in children and adolescents? Pediatr Infect Dis J December 2021;40(12):e482−7.

[19] Ludvigsson JF. Case report and systematic review suggest that children may experience similar long-term effects to adults after clinical COVID-19. Acta Paediatr March 2021;110(3):914−21.

[20] Osmanov IM, Spiridonova E, Bobkova P, Gamirova A, Shikhaleva A, Andreeva M, et al. Risk factors for long covid in previously hospitalised children using the ISARIC Global follow-up protocol: a prospective cohort study. Eur Respir J July 2021;1:2101341.

[21] Denina M, Pruccoli G, Scolfaro C, Mignone F, Zoppo M, Giraudo I, et al. Sequelae of COVID-19 in hospitalized children: a 4-months follow-up. Pediatr Infect Dis J December 2020;39(12):e458−9.

[22] Bharat A, Querrey M, Markov NS, Kim S, Kurihara C, Garza-Castillon R, et al. Lung transplantation for patients with severe COVID-19. Sci Transl Med December 16, 2020;12(574):eabe4282.

[23] Aiello A, Farzaneh F, Candore G, Caruso C, Davinelli S, Gambino CM, et al. Immunosenescence and its hallmarks: how to oppose aging strategically? A review of potential options for therapeutic intervention. Front Immunol 2019;10:2247.

[24] Kaseda ET, Levine AJ. Post-traumatic stress disorder: a differential diagnostic consideration for COVID-19 survivors. Clin Neuropsychol November 2020;34(7−8):1498−514.

[25] Kikkenborg Berg S, Dam Nielsen S, Nygaard U, et al. Long COVID symptoms in SARS-CoV-2-positive adolescents and matched controls (LongCOVIDKidsDK): a national, cross-sectional study [published online ahead of print, 2022 Feb 7]. Lancet Child Adolesc Health 2022. S2352-4642(22)00004-9.

[26] Interim Guidance on Supporting the Emotional and Behavioral Health Needs of Children, Adolescents, and Families During the COVID-19 Pandemic [Internet]. [cited 2021 Apr 20]. Available from: http://services.aap.org/en/pages/2019-novel-coronavirus-covid-19-infections/clinical-guidance/interim-guidance-on-supporting-the-emotional-and-behavioral-health-needs-of-children-adolescents-and-families-during-the-covid-19-pandemic/.

[27] Tenforde MW. Symptom duration and risk factors for delayed return to usual health among outpatients with COVID-19 in a multistate health care systems network — United States, March−June 2020. MMWR Morb Mortal Wkly Rep [Internet] 2020 [cited 2021 Dec 30];69. Available from: https://www.cdc.gov/mmwr/volumes/69/wr/mm6930e1.htm.

[28] Pinto AJ, Dunstan DW, Owen N, Bonfá E, Gualano B. Combating physical inactivity during the COVID-19 pandemic. Nat Rev Rheumatol July 2020;16(7):347−8.

[29] Marques IG, Astley C, Sieczkowska SM, Iraha AY, Franco TC, Smaira FI, et al. Lessons learned from a home-based exercise program for adolescents with pre-existing chronic diseases during the COVID-19 quarantine in Brazil. Clinics (Sao Paulo). 2021;76:e2655.

[30] American Academy of Pediatrics. Post-COVID-19 Conditions in Children and Adolescents [Internet]. https://www.aap.org. [cited 2021 Dec 30]. Available from: http://www.aap.org/en/pages/2019-novel-coronavirus-covid-19-infections/clinical-guidance/post-covid-19-conditions-in-children-and-adolescents/.

[31] Antonelli M, Penfold RS, Merino J, Sudre CH, Molteni E, Berry S, et al. Risk factors and disease profile of post-vaccination SARS-CoV-2 infection in UK users of the COVID Symptom Study app: a prospective, community-based, nested, case-control study. Lancet Infect Dis January 2022;22(1):43−55.

Index

'*Note:* Page numbers followed by "f" indicate figures "t" indicate tables and "b" indicate boxes.'

O

P

R

S

T

U

V

Printed in the United States
by Baker & Taylor Publisher Services